ICE BOUND

THE AUSTRALIAN STORY OF ANTARCTICA

ICE BOUND

THE AUSTRALIAN STORY OF ANTARCTICA

Joy McCann

Foreword by Tim Jarvis AM

NLA PUBLISHING

Foreword

Australia and Antarctica are old friends, with a relationship stretching back 85 million years to when we were joined at the hip as part of Gondwanaland and were both, improbably, home to marsupials.

Antarctica has always had an allure for us. Perhaps we are drawn by the mystique of a larger and more climatically challenging cold version of our own vast hot country. Perhaps it is the challenge of trying to unlock its secrets—one that inspired a generation of heroic-era polar explorers to head south to fill the blanks on our maps. Sir Douglas Mawson, Sir Hubert Wilkins, Frank Hurley, Thomas Griffith-Taylor, Frank Wild, John King Davis, Carsten Borchgrevink and Louis Bernacchi, to name just a few, endured physical privation and committed feats of great bravery in the process.

Mawson's leadership of the combined British, Australian, New Zealand Antarctic Research Expeditions from 1929 to 1931 secured the largest chunk of the Antarctic continent as Australian Antarctic Territory—a 42% claim that exists to this day. This proud legacy of exploration continues across a wide range of scientific disciplines, with Australian scientists leading the charge on the Antarctic continent in many fields of endeavour. And the gender imbalance, too, continues to steadily improve. The first female scientists, Susan Ingham, Isobel Bennett and Hope McPherson, went to the subantarctic in 1959; in 2021, the Australian Antarctic Division appointed Professor Nicole Webster as Chief Scientist, the second woman to hold the position.

This sense of history, combined with a privileged position in Antarctica, has led us to feel protective of the continent, with Australia assuming a prominent role in its defence. We were a key architect of and early signatory to both the 1959 Antarctic Treaty and the ground-breaking Madrid Protocol, which led to a ban on mining in Antarctica and to the continent being designated a 'natural reserve, devoted to peace and science'. More recently, Australia was lead negotiator of the Convention on the Conservation of Antarctic Marine Living Resources, which protects Antarctic marine life and pledges to create a network of marine parks.

This book is a timely and engaging reminder of this history and connection but, more importantly, Dr McCann captures the experience of living and working on the ice, of losing fellow expeditioners to cold, hunger and exhaustion, of hauling a sledge through ridges of sastrugi, of witnessing the effects of climate change in real time as icebergs calve from the edges of ice shelves. Although Australia and Antarctica physically parted ways 30 million years ago, as *Ice Bound* shows, the relationship and ties that bind us are as strong as ever—perhaps even stronger than the physical connection of old.

Tim Jarvis AM
Adventurer and environmental scientist

Contents

Preface

Antarctica is a paradox. This southern polar region is so inhospitable and so remote from civilisation that it seems to have little relevance for human life. Yet the chill of its polar atmosphere and the ebb and flow of its sea ice are intimately connected to the circulation of the world's oceans and atmosphere. For Australians, the Antarctic continent and its circumpolar Southern Ocean exert a profound influence on our island continent, shaping not only our weather but also our sense of place and history. As Prime Minister Robert Menzies noted, as he welcomed delegates to the first Antarctic Treaty Consultative Meeting at Australia's Parliament House in 1961, Australians 'have a deep sense of neighbourhood about the Antarctic'.[1]

Much has been written about Australia's connections—both imaginative and real—with the Antarctic region. Stories of ocean, ice, winds, wildlife, exploration, adventure, exploitation, conservation, science and governance abound. Tom Griffiths has observed how strange it seems that settler Australians, as inhabitants of an arid continent, were drawn to colonise a continent of ice, 'although', he adds, 'Australians *do* know something about deserts'.[2] Indeed, over the past century, writers, artists and scholars have often been struck by the similarities between the Australian outback and the Antarctic 'desert' as they pondered its significance for Australians.

This book shows some of the myriad ways in which Australians have made sense of the Antarctic region. It offers a selection of stories drawn from many different sources: literature and art, scientific research and government reports, places and objects, diaries, images and memories. It traverses time and space, taking the reader on a journey south, across the Southern Ocean, onto the beaches of Australia's subantarctic island territories and into the ice continent itself. This book also seeks to locate these individual stories of the high southern latitudes within the larger histories of polar exploration, national sovereignty, scientific inquiry, exploitation, conservation and adaptation. It concludes with a detailed list of sources and suggested reading for those who may wish to delve further into the rich collections of books, documents, oral histories, images, films, maps and manuscripts and explore some of the connections between Australia and Antarctica.

Powerless, one was in the spell of an all-enfolding wonder. The vast, solitary snow-land, cold-white under the sparkling star-gems; lustrous in the radiance of the southern lights; furrowed beneath the icy sweep of the wind. We had come to probe its mystery, we had hoped to reduce it to terms of science, but there was always the 'indefinable' which held aloof, yet riveted our souls.

SIR DOUGLAS MAWSON[1]

Continent

← Wind-etched snow surface on Law Dome.

ustralia and Antarctica are the only continents to lie entirely within the Southern Hemisphere. They share the dubious honour of being the two driest continents on Earth, but they seem utterly different in every other respect. Australia's climate ranges from tropical to temperate, with regions spanning alpine ecosystems to grasslands and deserts; Antarctica—nearly twice the size of Australia—is blanketed almost entirely by a permanent ice sheet. First Nations peoples have lived in Australia for at least 65,000 years; Antarctica is the only continent to have no enduring human population. Nevertheless, these two landmasses are deeply connected across time and space. Separated by the Southern Ocean, southern Australia is closer to Antarctica than to the countries of South-East Asia. The nearest research station on the Antarctic continent (the French Antarctic research station Dumont d'Urville) lies about 2,700 kilometres south of Hobart: less than the distance by air between Perth and Sydney (3,920 kilometres).[2]

↓ In 2000, B-15—the biggest iceberg on record at the time—calved from the Ross Ice Shelf, the world's largest body of floating ice.

Antarctica: the ice continent

The Antarctic continent is almost entirely covered by ice, much of which is more than two kilometres thick (only 0.4 per cent of the continent is ice free). The ice sheet formed over hundreds of thousands of years as a result of snow falling and compacting into thick layers that hold about 70 per cent of Earth's fresh water. Under its own sheer weight, the ice is constantly on the move, flowing slowly downhill from the interior to the coast. Nearer the coast, ice is propelled along great rivers known as glaciers and coalesces in places to form ice shelves that push out into the surrounding ocean. These permanent floating ice shelves maintain a dynamic stability, expanding as more glacial ice builds up behind them and shrinking as icebergs calve from their edges. Some ice shelves extend for many kilometres along the coastline and, in the more protected bays, can survive for thousands of years. The Ross Ice Shelf at the head of the Ross Sea is the world's largest body of floating ice.

The winter air temperature in the highest parts of the ice sheet averages −60°C. In 1983, the lowest surface temperature ever recorded at ground level at that time was −89.2°C, taken at the site of the Soviet station at Vostok.[3] These elevated places in the inland plateau receive little precipitation, with snowfall equivalent to just five centimetres of rainfall a year. The conditions are perfect, however, for the formation of intensely cold, dense air to flow down from the elevated areas, creating the notorious katabatic winds that accelerate downhill and blow over the coastal areas before they dissipate offshore. Blizzards, too, are a defining feature of the Antarctic ice sheet, where freezing gale-force winds can blow for hours or days on end.[4]

The ice sheet dominates the Antarctic continent, but small amounts of rock protrude above the surface in the form of rocky coastal outcrops, mountain ranges, nunataks (isolated rocky peaks surrounded by ice) and massifs. Areas of rock exposed above the ice are a rarity and have assumed a privileged place in this ice continent. All terrestrial life in Antarctica relies on these oases to survive. A few hardy species of lichen and algae cling to rock among the ice and snow, while specialised grasses and mosses survive in the few ice-free places closer to the coastline. Human colonisation has also depended on these rocky outcrops. Along the coastline, they provide access for ships and a stable foundation on which to anchor buildings against the ravages of this blizzard-prone land. Inland, the peaks of mountain ranges and rare ice-free depressions known as 'dry valleys' offer a window into Antarctica's deep geological past.[5]

Most areas of exposed rock are found on the Antarctic Peninsula (the warmest part of the continent) and along the Transantarctic Mountains. They are the highest mountains in Antarctica—rising to 4,528 metres at their highest point—although only their peaks can be seen above the ice sheet. They are also one of the longest mountain ranges in the world, stretching for 3,200 kilometres from Victoria Land to the Weddell Sea.[6] These mountains divide the continent into two geologically distinct parts: East Antarctica (or Greater Antarctica) and West Antarctica. East Antarctica is considerably larger and higher than the western portion and the Australian Antarctic Territory lies wholly within it.

↑ An aerial view of a glacier in the Transantarctic Mountains.

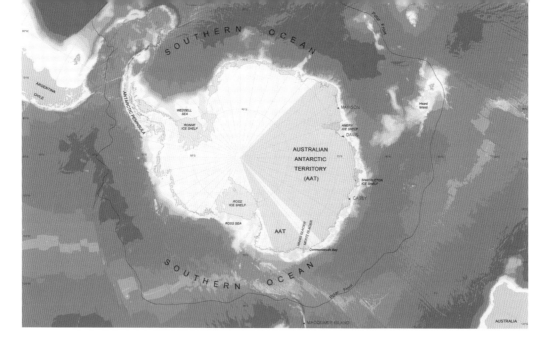

↑ The Australian Antarctic Territory covers 42% of Antarctica (nearly 80% of Australia's total area). As a signatory to the Antarctic Treaty, Australia is committed to international scientific cooperation and ensuring that the continent is used exclusively for peaceful purposes.

The geology of East Antarctica is not well known, but it appears to contain the bulk of the Antarctic ice sheet cradled in a deep depression. Below the ice sheet lies the East Antarctic Shield: ancient igneous and metamorphic rocks estimated to be up to four billion years old. They are among the oldest known rocks on the planet, and just a few hundred million years younger than Earth itself. This ancient continental rock shield protrudes above the ice sheet in the form of coastal outcrops, mountain ranges or isolated peaks, including places now known as the Bunger Hills and Vestfold Hills near Australia's Davis research station, the Prince Charles Mountains inland of Australia's Mawson research station, and a portion of the Transantarctic Mountains in the eastern sector of the Australian Antarctic Territory. West Antarctica is divided from the larger eastern portion by the Transantarctic Mountains. It comprises a patchwork of land and ice separated by deep ocean water. Indeed, the West Antarctic Ice Sheet is regarded as a floating ice mass, anchored by terrestrial landmasses that are still volcanically active. This portion of the continent includes the Antarctic Peninsula and the highest mountain peak in Antarctica: the Vinson Massif, measuring 4,892 metres above sea level.[7]

The Australian landscape may be vastly different to that of Antarctica, but the two continents share an ancient lineage: both were created in deep geological time from the

→ Discoveries of fossilised plants and animals have enabled scientists to understand how the modern Antarctic environment evolved.

↓ *Cryolophosaurus* (meaning 'frozen crested lizard') is one of the earliest known meat-eating dinosaurs. It roamed Antarctica during the Early Jurassic epoch around 170 million years ago.

supercontinent known as Gondwana. When Gondwana began to break apart around 180 million years ago, the resulting forces of plate tectonics began the slow process of severing the ancient continental bond forever. The plate carrying the future landmasses of Africa, South America and India gradually drifted towards the equator, while Australia, the last fragment to separate from Antarctica, embarked on its own voyage northward about 85 million years ago. The plate moved slowly over the next 50 million years, until it too was breached by the mighty Antarctic Circumpolar Current of the Southern Ocean, giving birth to the great southern Gondwanan cousins. There was no ice sheet covering Antarctica at this time, and dinosaurs, amphibians and marsupials migrated freely across the Gondwanan landscape, untroubled by oceanic barriers. As the Antarctic landmass continued on its southward journey, it became colder and more hostile to the plant and animal life that remained on this doomed life raft. The interior of the continent still bears traces of this ancient ancestry, including some of the oldest rocks, formed more than 3 billion years ago, at Enderby Land in the Australian Antarctic Territory, and the fossilised remains of temperate plants and animals, including dinosaurs and marsupials.[8]

Sensing the south

These physical connections between Australia and Antarctica are a powerful reminder of Earth's geological history. The two continents have cultural connections too. Creation stories from lutruwita (Tasmania) tell of great winds and seas that lashed at the island and icebergs that floated from the south.[9] European trading vessels, bound for the East Indies (Indonesia), made use of these powerful westerly winds to propel their ships eastward across the vast expanse of ocean. From the sixteenth century, those who ventured further south beyond the Cape of Good Hope and Cape Horn gave names to these winds of the high southern latitudes—the 'Roaring Forties', the 'Furious Fifties' and the 'Screaming Sixties'—invoking the sound of the wind blowing through the rigging of their ships. When the ships of northern empires followed in their wake, bound for the coastlines and islands of Australia, convicts and colonists alike took their chances with the tumultuous winds of the Southern Ocean. These winds have their origins in the uneven air pressures over Antarctica and are whipped up into a fury as they travel over the vast expanses of ocean with no landmass to interrupt their flow. Together, the circumpolar winds and ocean currents conspire to create some of the stormiest sailing conditions on Earth. As shipwrecks littered along Australia's southern coastlines attest, a miscalculation of longitude could spell disaster. But a ship's fate in these southern latitudes would also be determined by the vast air masses emanating from Antarctica.

Settler Australians sensed Antarctica long before they knew of its existence. As the historian Tom Griffiths put it, 'they felt its breath'.[10]

In 1861, as the Australian gold rushes gathered pace, Matthew Fontaine Maury (then Superintendent of the United States Hydrographic Office) advised the British Admiralty to establish a network of meteorological stations across the high southern

← Augustus Earle depicted the turbulent conditions off the Cape of Good Hope in 1824 after his rescue from Tristan da Cunha by the *Admiral Cockburn* which was bound for Van Diemen's Land.

↑ The 'clipper route' became popular in the mid-19th century as the fastest way to sail between England and Australia in the thrall of the Roaring Forties, but shipwrecks were common in the stormy Southern Ocean.

latitudes. The resulting observations, he noted, would be of immense practical value for both navigation and science in the Southern Hemisphere.[11] Great Britain had colonised one of the driest continents on Earth, and understanding the weather patterns of these ocean latitudes would become a matter of life and death.[12] By 1893, as large swathes of inland Australia felt the grip of a prolonged and devastating drought, the government astronomer Henry Chamberlain Russell began advocating for systematic meteorological observations in order to understand the influence of the Antarctic region on the movement of pressure systems over Australia.[13] Then, in 1901, as the colonies prepared to celebrate Federation in Melbourne, Queensland's government meteorologist Clement Wragge sent word to expect 'fierce westerly squalls with driving rain' on the day. 'The Federal Parliament,' he predicted, 'will be opened amid the blustering grandeur of a blow from Antarctica.'[14] It was becoming clear that the temperamental polar airs could play havoc with Australia's weather patterns. The problem, observed geographer John Walter Gregory, was that Australians lacked an understanding of the physical nature of Antarctica and its effects on climatic conditions to its north: 'any variation in their condition inevitably affects our natural prosperity and welfare'.[15]

When Robert Falcon Scott's *Terra Nova* expedition sailed south to Antarctica in 1910, the British-born Australian geographer Thomas Griffith Taylor was on board.[16] 'Griff' had trained in mining and metallurgy at Sydney University with Professor (later Sir) Tannatt William Edgeworth David and, as expedition geologist, he would lead two sledging journeys, make extensive studies of glaciation processes, and produce the first geological maps of the continent. He also held the position of Physiographer with the Commonwealth Weather Service in Australia, assisting the expedition meteorologists George Simpson and Charles Wright to conduct balloon experiments designed to record the movements of the upper air over Antarctica. Significantly, the data he collected would be crucial in developing a better understanding of how Antarctica's atmosphere influenced the weather systems in the Southern Hemisphere, especially Australia. During the expedition, Griff earned a reputation as a dedicated scientist with a prodigious talent for writing and expressing his ideas and observations.[17] In his memoir, Apsley Cherry-Garrard described Griff's larger-than-life presence during the expedition:

His gaunt, untamed appearance was atoned for by a halo of good-fellowship which hovered about his head. I am sure he must have been an untidy person to have in your tent: I feel equally sure that his tent-mates would have been sorry to lose him. His gear took up more room than was strictly his share, and his mind also filled up a considerable amount of space. He always bulked large, and when he returned to the Australian Government, which had lent him for the first two sledging seasons, he left a noticeable gap in our company.[18]

Powerful weather systems were not the only harbinger of what lay to the south of Australia. Auroral displays, known to colonists by the Latin name aurora australis or 'southern lights', were sometimes seen in the night skies above southern Australia and New Zealand during major solar eruptions.[19] Aurorae occur in both hemispheres

↓ Named for the Roman goddess of the dawn, the aurora australis shimmers across the night skies, giving Australians a glimpse of the powerful forces at work in the southern polar region.

when gases in the atmosphere collide with charged particles travelling along Earth's magnetic fields around the poles, creating shimmering waves of colour—predominantly green, red and violet—in the night sky (the phenomenon is called aurora borealis in the Northern Hemisphere). Sightings of the aurora australis hold an important place in the oral traditions of First Nations communities across southern Australia. Aurorae are commonly a reddish colour in Australian skies, and some groups associated their appearance with fire in the spirit world. In other instances, the appearance of the aurora australis was feared, either as an omen of spirit beings with supernatural powers or as a harbinger of death.[20]

Australian colonists attributed their own meanings to aurorae, with articles appearing in local newspapers noting their appearance, location and timing in relation to terrestrial magnetism—a subject of keen interest in the early nineteenth century as scientists sought to understand how Earth's magnetic field functioned. In 1835, for example, an article in *The Hobart Town Courier* described how:

> On Wednesday evening a most splendid aurora australis was visible in this hemisphere. The heavens were brilliantly illuminated, the phenomenon beginning to appear in the west, and gradually extending southward till nearly the entire semi horizon was enlightened, the coruscations occasionally drawing towards a focus at the zenith, or at least 80 degrees high. It lasted several hours. It was nearly as light at Hobart Town as during the first quarter of the moon.[21]

Two decades later, *The Argus* reported a particularly striking spectacle in Melbourne following a magnetic disturbance detected at Melbourne's Flagstaff Observatory: 'soon after sun-set an aurora was displayed such as has seldom been witnessed in the Southern Hemisphere'.[22]

And so the Antarctic region was taking shape in the Australian imagination, long before humans ventured into the high southern latitudes. Through their stories about ice and wind, ocean and wildlife, light and darkness, Australians found ways to make sense of the mysterious realm to their south. In the process, they created powerful cultural connections between these two Gondwanan cousins that endure to this day.

↑ Sleeping tents for an ice drilling field party at Aurora Basin North Camp,
inland from Australia's Casey research station, in December 2013.

Gradually the swell subsided, smoothed by the weight of ice.
The tranquillity of the water heightened the superb effects
of this glacial world. Majestic tabular bergs whose crevices
exhaled a vaporous azure; lofty spires, radiant turrets and
splendid castles; honeycombed masses illumined by pale
green light within whose fairy labyrinths the water washed
and gurgled. Seals and penguins on magic gondolas were
the silent denizens of this dreamy Venice. In the soft glamour
of the midsummer midnight sun, we were possessed by a
rapturous wonder—the rare thrill of unreality.

SIR DOUGLAS MAWSON[1]

Ice

← Icebergs calved from glaciers or ice shelves often reveal striking shapes and colours as they are eroded by wind and waves.

Since the second century CE, geographers and philosophers were intrigued by the idea that a great landmass lay at the southernmost end of Earth. The Classical Greek astronomer and geographer Claudius Ptolemy sowed the seeds of an enduring fascination with the southern polar region when he speculated that a great landmass must exist south of the equator in order to counterbalance the great landmass in the Northern Hemisphere.[2] This imagined geography of a spherical Earth with two poles took hold, persisting into the fifteenth century and converging with new information brought back by navigators from their forays into the oceans south of the equator. Renaissance cartographers responded with a variety of fantastic visual motifs to represent these new constructions of the world's southernmost geography, from an icy sea that surrounded the South Pole to a vast continent, *Terra Australis nondum cognita* (unknown southern land), centred on the pole and extending almost to the equator.[3]

Towards the South Pole

When British naval lieutenant James Cook embarked on his first voyage into the Southern Hemisphere aboard the barque *Endeavour* in 1768, his mission was to observe the transit of Venus in Tahiti. He was also given a second set of orders from the British Admiralty, handed to him under seal, to ascertain whether the mythical southern continent existed and, if so, to claim it in the name of the king of Great Britain.[4] Cook returned to Britain after three years without conclusive proof that the unknown southern land existed, so a year later he embarked on a second voyage into the high southern latitudes, in command of the *Resolution* and *Adventure*. At latitude 48° S, the air and sea temperatures suddenly plummeted, and they encountered 'a small gale, thick foggy weather, with much snow; thermometer from 32 to 27; so that our sails and rigging were all hung with icicles'.[5] Most of the animals that were taken on board at the Cape of Good Hope as food for the long voyage perished from the cold. At latitude 51° S, the crews sighted their first icebergs and were soon surrounded by an 'immense field of Ice'. They turned north then circled south again, forever surrounded by floating ice. At latitude 61° S, Cook ordered three boats into the icefield to collect loose pieces that could be melted

→ The *Resolution* at anchor
among icebergs, drawn by William
Hodges, an artist who accompanied
James Cook on his second voyage
of exploration in search of the
'unknown southern land'.

down for fresh water, then,
on 17 January 1773 at latitude
66° S, the ships crossed the
Antarctic Circle, the first
European vessels known to
have done so.

Enveloped in a thick blanket
of fog, the captains found it exceedingly difficult to navigate among the 'ice islands'.
Somewhere in the vicinity of the Kerguelen Islands (Îles Kerguelen, claimed by France
in 1772), the *Adventure*, under the command of Tobias Furneaux, lost contact with the
Resolution and was forced to sail on alone to Van Diemen's Land (Tasmania) before
heading to the prearranged meeting point at Queen Charlotte Sound on New Zealand's
South Island. Cook, meanwhile, sailed southeast, tracing the edge of the icefield, before
also making for Queen Charlotte Sound where the two ships were reunited in May 1773.
After several months exploring the central Pacific, the two ships were again separated
in a storm off New Zealand. This time Furneaux decided to return to Britain, arriving
in July 1774—a year before Cook, who had turned the *Resolution* southward to resume
his search for the southern continent.

Over the next four months, Cook steered the *Resolution* back and forth across the
Southern Ocean, circumnavigating the Antarctic continent and passing close to it
without ever sighting land. He crossed the Antarctic Circle for the second time on
20 December 1773, but the ice and cold forced him to turn north to warmer waters. He
resumed his search once more and, on 26 January 1774, he crossed the Antarctic Circle
for a third time before reaching the southernmost point of his voyage at latitude 71°10' S.
Finding his ship's path blocked by a vast field of solid pack ice, he wrote:

I will not say it was impossible anywhere to get farther to the south; but the attempting it would have been a dangerous and rash enterprise, and what, I believe, no man in my situation would have thought of. It was, indeed, my opinion, as well as the opinion of most on board, that this ice extended quite to the pole, or perhaps joined on some land, to which it had been fixed from the earliest time; and that it is here, that is to the south of this parallel, where all the ice we find scattered up and down to the north, is first formed, and afterwards broken off by gales of wind, or causes, and brought to the north by the currents, which we always found to set in that direction in the high latitudes.[6]

→ World maps showing the tracks of James Cook's three voyages of exploration were first issued in 1784. This 1838 map highlighted the extent of Cook's discoveries in both hemispheres as well as subsequent British expeditions to the 'north polar seas'.

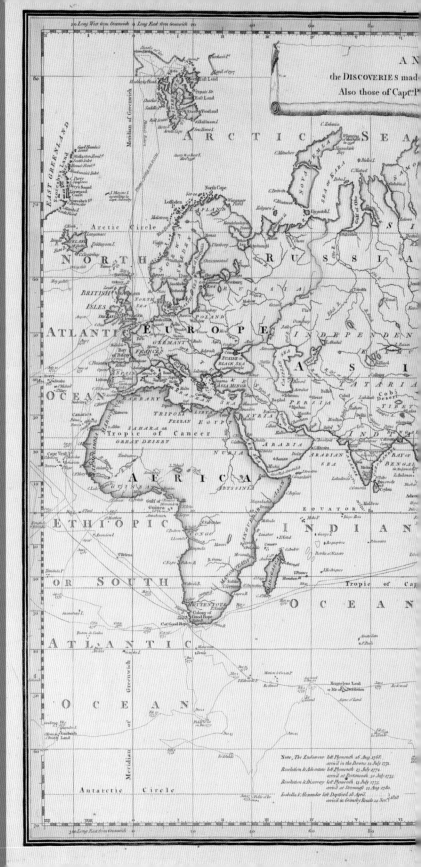

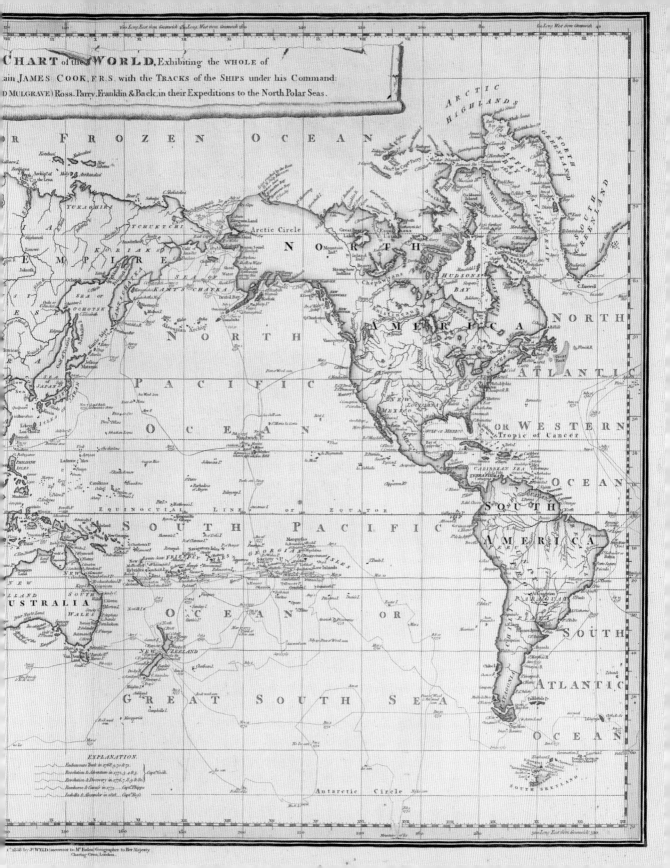

The mythical southern continent would remain elusive—for the moment at least—although the publication of Cook's reports in 1777 offered the first tantalising glimpses, not of an El Dorado at the South Pole, but of an abundance of whales and fur seals around the waters of the subantarctic islands.

An ocean of riches

Animal fur attracted high prices in the clothing markets of Canton (Guangzhou), London and New York, and the soft underfur of southern fur seals was in high demand. The process of harvesting, however, was a ruthless and brutal business. Sealing captains would dispatch a gang of sealers to an island for weeks or months at a time. Their sole purpose was to slaughter and skin every seal they found until whole beaches were empty. The gangs operated with impunity on these remote subantarctic islands, their activities unregulated and secretive, and little is known about them. They were often the first Europeans to set foot on many of these isolated shores, camping in caves or constructing crude huts from stone, timber or the ribs of a whale covered with seal skins to shelter from the elements.[7] Archaeologists and historians are still unearthing the physical remains of these remote beach camp sites and burial sites.

By the 1820s, the fur seal populations of the subantarctic islands had been reduced from an estimated one to two million creatures to just a few hundred.[8] By then, hunters were turning their attentions to the southern elephant seals (*Mirounga leonina*) and baleen whales of the Southern Ocean. Baleen whales, like the mighty blue (*Balaenoptera musculus*), southern right (*Eubalaena australis*) and humpback (*Megaptera novaeangliae*)

↓ When this seal hunt took place during Jules Dumont d'Urville's Antarctic expedition (1837–1840), unregulated hunting had already devastated the once-abundant populations of fur seals in the Southern Ocean.

whales, migrate each year along ancient sea routes between their breeding grounds in the warmer waters and sheltered bays of southern Australia and their feeding grounds in the cold, fertile waters around the Antarctic continent and subantarctic islands. The oil derived from these blubber-rich marine mammals was in high demand in the growing cities and towns of the Northern Hemisphere, lighting the streets and lubricating the engines of the industrial revolution as it swept across Europe and North America.

Enterprising Australian colonists also saw the potential rewards to be reaped from whaling, establishing processing stations along the extensive southern coastlines and islands of Australia. Using small whale-catcher boats to hunt the whales in bays, they would tow the carcass to shore where they would flense blubber from the massive bodies and melt it down to extract the precious oil. The first Australian shore-based station was established at Ralphs Bay near Hobart Town in 1806 and, by 1841, there were an estimated 58 bay whaling stations in southern Van Diemen's Land alone.[9] Bay whaling had become one of the earliest and most lucrative industries in the Australian colonies, attracting enterprising individuals like the sealer 'Captain' James Kelly, who served as Hobart's harbour master and pilot from 1818 until retiring to take

up whaling in 1829. As the populations of elephant seals and whales in the shallower coastal waters of southern Australia rapidly declined, the hunters began venturing further south in search of their prey. By then, sealing and whaling vessels bound for the southern polar region had become a regular sight in Australia's southern ports, undertaking boat repairs and providing shore leave for their crews. The Derwent River estuary in Hobart Town was a popular resupply point for ships engaged in this profitable new industry, offering a sheltered port for whaling vessels as they voyaged south into the circumpolar storm track of the Southern Ocean. Between 1816 and 1823 alone, some 300 ships were listed as having anchored in the Derwent, almost half of which were sealing and whaling vessels from North America.[10]

↗ The seasonal abundance of whales along Australia's southern coastlines created a lucrative coastal whaling industry in the early 19th century.

⇢ By 1850, there were few southern right whales left in Hobart's Derwent River and whalers turned to hunting sperm whales offshore.

Antarctica's forests

Botanist Joseph Hooker, who sailed with James Clark Ross's Antarctic expedition in 1839–1843, was the first to recognise a similarity between plant species in the continents and islands of the high southern latitudes. He speculated that they may have originated from a single landmass at the South Pole, and this idea gained momentum when subsequent Antarctic expeditions discovered evidence of fossilised wood and plants on the Antarctic Peninsula and nearby islands.[11]

One of the most important of these discoveries was made by scientist, physician and artist Edward Wilson, during Robert Falcon Scott's ill-fated attempt to lead the first team to reach the South Pole in 1911. When rescuers located the frozen bodies of Scott, Wilson and Henry Bowers in their tent the following spring, they also found rock samples bearing fragments of fossilised leaf that the men had hauled from the base of Beardmore Glacier. The British palaeobotanist A.C. Seward later identified the fossilised remains as belonging to the species *Glossopteris indica*, an ancient fernlike plant known from fossils found in India. The discovery of *Glossopteris* in Antarctica provided a crucial piece of evidence to support the theory—first proposed by the Austrian geologist Eduard Suess—that land bridges once connected South America, Africa, India, Australia and Antarctica, forming the ancient supercontinent of Gondwana.

Subsequent fossil discoveries have provided further evidence that the frozen white desert once supported a flourishing temperate rainforest. It seems that life existed in Antarctica similar to other southern continents until at least 55 million years ago, although scientists are uncertain about exactly when Antarctica's forests died out. The discovery of fossilised remains 400 kilometres from the South Pole of *Nothofagus beardmorensis*, an extinct relative of the southern beech tree that still grows in modern Tasmania, South America, New Zealand, New Guinea and New Caledonia, offers tantalising new evidence suggesting that the Antarctic ice sheet may be much younger than previously thought: perhaps as young as two to three million years old.[12]

→ The myrtle beech (*Nothofagus cunninghamii*) is one of three native Australian *Nothofagus* species. Fossilised leaves found in Antarctica reveal the continent once supported a temperate rainforest.

The lure of the ice

Expedition and whaling ships bound for the far south were becoming a familiar sight to the people of Sydney, Melbourne and Hobart and invariably drew a curious crowd.[13] In April 1820, *The Sydney Gazette and New South Wales Advertiser* reported that Captain Fabian Gottlieb von Bellingshausen of the Imperial Russian Navy dropped anchor in Port Jackson (Sydney Harbour), having completed a circumnavigation of the Antarctic Circle aboard the *Vostok*.[14] A decade later, John Biscoe 'was obliged to put in at Hobart Town in consequence of sickness among the crew' after several men perished at sea during the voyage to Antarctica aboard the whaling brig *Tula*. Biscoe worked for the Enderby Brothers of London, a whaling and sealing company that combined hunting activities with exploration. According to (Sir) Douglas Mawson, Biscoe was 'the first ever to catch sight of any portion of the Antarctic continent', when he sighted a barren, ice-capped plateau rising to a coastal mountain range during his voyage. Biscoe named it Enderby Land after his employers.[15] (It was subsequently claimed as part of the Australian Antarctic Territory.)

Jules Dumont d'Urville's French scientific expedition aboard the *Astrolabe* and *Zelée* spent 15 days in Hobart Town in December 1839, with several sick crew members who became ill during 'a long and painful voyage of discovery to the south of Cape Horn' and across the Pacific.[16] Another Antarctic expedition to receive the hospitality of Hobart Town arrived with the *Erebus* and *Terror* on 16 August 1840, under the command of James Clark Ross. By the time Ross returned to Hobart on 7 April 1841, his polar expedition had become a *cause célèbre* in the small colonial outpost. 'Last evening', the Hobart *Courier* exclaimed, 'the much talked of and anxiously looked for spectacle of the Antarctic expedition was performed before a densely crowded house of all classes and sizes.'[17] It was probably the first play ever performed about the Antarctic.

The number of Antarctic-bound ships in southern Australian ports slowed dramatically after Ross's expedition, although the mysteries of the far south continued to intrigue those in colonial scientific circles. One of the first efforts to promote Australia's scientific interests in the polar region came from the Royal Society of Victoria, which established an Australian Antarctic Exploration Committee in 1886 to convince the colonial

↑ The Royal Society established a network of magnetic observatories across
the British Empire, such as Rossbank, Hobart, established by James Clark Ross.

governments to support an expedition, not just as a means to claim 'this vast unclaimed
and unnamed territory', but also to pursue questions of scientific and economic
importance to Australia.[18] These expeditions were inspired in part by the quest to
understand the nature of Earth's magnetic forces. Indeed, the science of magnetism
in the polar regions was a popular subject of discussion in colonial newspapers of the
day. The campaign, known as the 'magnetic crusade', was led by scientists including
Edward Sabine and John Herschel, who advocated a global network of observatories
to collect data on Earth's magnetic field, including at the poles. By 1840, there were
more than 30 observatories around the world. The crusade reflected a preoccupation in
Victorian-era science with measuring the cosmos and determining its laws. Australia's
first observatory, established in 1861–1863, can still be seen in Melbourne's Royal
Botanic Gardens.[19]

With growing international interest in the geological, meteorological and geomagnetic characteristics of the polar regions, several nations agreed to establish scientific bases in the Arctic in preparation for the First International Polar Year, to be held in 1882–1883. Antarctica was not far behind, and some colonists were keen to promote Australia's interests in such research. In 1890, Alfred Joseph Taylor, a Hobart librarian, publicised plans for a jointly funded Swedish–Australian expedition, drawing attention to the importance of investigating the 'unexplored wastes beyond the Antarctic Circle' in order to gain essential knowledge about the natural history of the planet as a whole.[20] By then, Antarctic exploration seemed a less daunting prospect with the development of larger steam-powered vessels that could penetrate floating pack ice, which had made earlier voyages so treacherous for sailing ships. Such developments also brought a renewed interest in commercial whaling in the waters around the ice-bound continent.

Two Norwegian-born immigrants were the first Australians to venture into Antarctic waters. In 1893, businessman Henrik Johan Bull, who had migrated to Melbourne in the 1880s, led an expedition in search of southern right whales aboard the three-masted barque *Antarctic*. The expedition was financed by the Norwegian shipping and whaling magnate Svend Foyn. When the *Antarctic* docked in Melbourne, Carsten Egeberg Borchgrevink, who was working in Australia as a surveyor and teacher of languages and natural science, seized the opportunity to pursue his passion for collecting rocks and lichens and signed up as a deckhand. The voyagers encountered their first large iceberg near the Kerguelen Islands, and then one of the Southern Ocean's notorious storms descended:

> *After a few days of this violent plunging, you get more accustomed to the complicated motions necessary to preserve a balance, and can faintly appreciate the wild beauty of the well-named 'roaring Forties'. The endless succession of green or gray-black mountainous billows, their breaking crests, which are blown into shreds by the squall, and carried through the rigging with a shrieking and howling as of loosened demons, the flying storm-rent clouds, and frequent mist and rain, make a picture supremely grand in its own way. With the addition of icebergs and darkness, however, I confess that a beauty of a milder type would*

→ Norwegian-born Australian businessman Henrik Johan Bull and surveyor and teacher Carsten Egeberg Borchgrevink claimed to have been the first humans to set foot on the Antarctic continent. The *Illustrated Australian News* featured this engraving depicting them landing at Cape Adare in 1895.

have been sufficient for me, who had never before in person realized the astounding violence of a gale in Southern latitudes, and the rapidity with which a calm will change into a hurricane, and the latter into a calm once more.[21]

It was almost midnight when the ship finally managed to navigate through sea ice and steam along the ice cliffs of the continent, the sun's rays reflecting through crystals of ice and snow. It was unlike any land they had ever seen. There were no indigenous peoples, no rivers and no forests. Instead, they found a vast, white landscape that was both awe-inspiring and deeply unsettling. 'What impresses visitors,' Bull wrote, 'is the utter desolation, the awesome, unearthly silence pervading the whole landscape.' The land had a 'deathlike calm and immobility', a stark contrast to the perpetual motion of the swirling ocean currents and ice circling the ship. A week later, Bull and Borchgrevink landed at Cape Adare in the Ross Sea. These travellers from the far north of the world later claimed to be the first humans to set foot on the Antarctic continent, although there were other claims.[22]

Southern Cross

After his return to Australia, Borchgrevink prepared a report on his observations from Antarctica for the Sixth International Geographical Congress, held in London in 1895. His report, together with his lectures and published articles, served to kindle public enthusiasm for Antarctic exploration and science, with the Congress agreeing that 'the exploration of the Antarctic Regions is the greatest piece of geographical exploration still to be undertaken'.[23] Within three years, with financial support from the British publisher

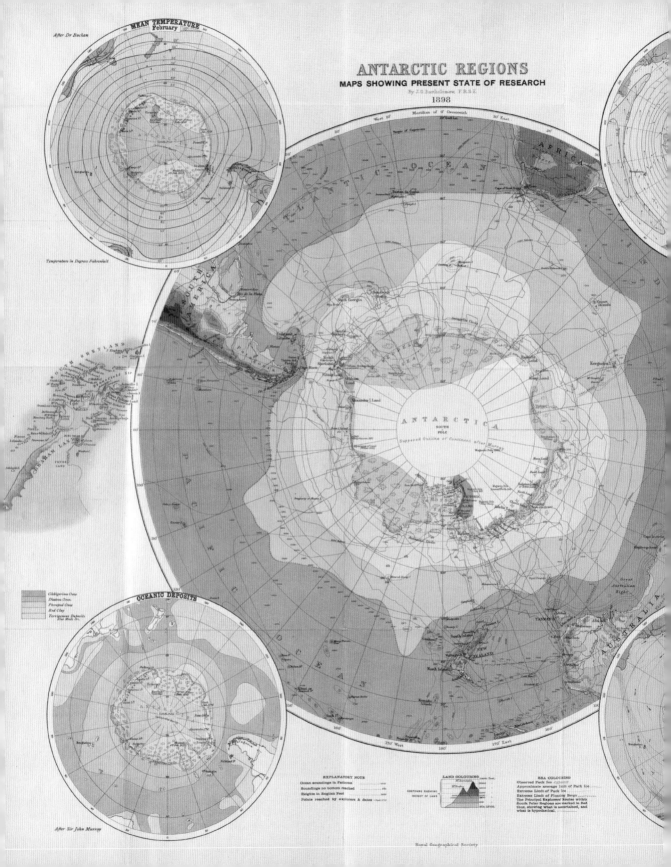

ANTARCTIC REGIONS
MAPS SHOWING PRESENT STATE OF RESEARCH
By J.G.Bartholomew, F.R.S.E.
1898

MEAN TEMPERATURE
February
After Dr Buchan

Temperature in Degrees Fahrenheit

OCEANIC DEPOSITS

Globigerina Ooze
Diatom Ooze
Pteropod Ooze
Red Clay
Terrigenous Deposits
Blue Muds &c.

After Sir John Murray

EXPLANATORY NOTE

Ocean soundings in Fathoms
Soundings no bottom reached
Heights in English Feet
Points reached by explorers & dates

LAND COLOURING

CONTOURS SHEWING
HEIGHT OF LAND

SEA LEVEL

SEA COLOURING

Observed Pack Ice
Approximate average limit of Pack Ice
Extreme Limit of Pack Ice
Extreme Limit of Floating Bergs
The Principal Explorers' Routes within
South Polar Regions are marked in Red
thus, showing what is ascertained, and
what is hypothetical.

← Fifty years after James Clark Ross's Antarctic expedition (1839–1843), advocates were calling for a new era of scientific and geographical exploration in the little-known Antarctic region. Clements Markham successfully argued for a British National Antarctic Expedition led by Robert Falcon Scott.

→ Visitors to Hobart's waterfront are reminded of the city's close Antarctic connections in this statue of the Tasmanian Louis Bernacchi, who served as physicist with Borchgrevink's British Antarctic (*Southern Cross*) Expedition (1898–1900) and Scott's British National Antarctic (*Discovery*) Expedition (1901–1904).

Sir George Newnes, Borchgrevink returned to Antarctica aboard the *Southern Cross* as the leader of a multinational expedition comprising Norwegian, Finnish, British and Australian men. Louis Charles Bernacchi, whose family had migrated to Tasmania from Belgium in 1884, had trained at Melbourne Observatory in the use of navigation and magnetic instruments, and Antarctica offered him the opportunity of a lifetime to study polar meteorology and terrestrial magnetism. Enthused by Borchgrevink's account of the *Antarctic* voyage, Bernacchi sailed for London to volunteer his skills as a physicist and astronomer.

By December 1898, Bernacchi was on his way back to the Southern Hemisphere aboard the *Southern Cross*. As the vessel lay at anchor in the Derwent River, Hobartians turned out in their thousands to welcome the expeditioners. Bernacchi later wrote: '[t]he whole town appeared to us to have but one occupation—entertaining the members of the Expedition'. As the ten expeditioners and 75 dogs finally set sail from Hobart, they were heading into an ocean that had long been traversed by sealers and whalers, to a continent that was still virtually unknown. It was to be the first expedition to spend winter on the Antarctic continent.

↑ J.W. Beattie, Tasmania's official photographer, captured this image of the *Southern Cross* in Hobart prior to its departure for Antarctica in 1898. This expedition marked the beginning of the 'heroic era' of Antarctic exploration, but its pioneering achievements were overshadowed by the success of Scott's *Discovery* expedition.

Borchgrevink's plan was to land at Cape Adare and build a hut that would provide shelter over the dark winter months and a base for studying polar meteorology, magnetism and zoology. On 30 December 1898, after the long voyage south from Hobart, the *Southern Cross* met the pack ice. Bernacchi wrote:

> *Within an hour of sighting the first piece, we were in the thick of it. It was truly a marvellous sight to one who had spent nearly the whole of his life in a semi-tropical climate, and mere words can in no way adequately express those first sensations ... We were no sooner well in the pack than the beautiful white ice-petrel (Pagodroma nivea) appeared. It is one of the most graceful birds on the face of the globe, and is never to be found far from the southern ice-fields ... Soon after the long polar night, when we were badly in want of fresh food, we endeavoured to eat some of these beautiful creatures. Alas! Although so fair to look upon, a closer acquaintance proved altogether undesirable, for the flesh was tough, dark, and utterly flavourless.*[24]

On 17 February 1899, the expedition arrived at Cape Adare in the midst of a storm. 'The storm raging with full fury,' Bernacchi wrote in his diary. '[S]uch a scene I have never before witnessed. The horror of it cannot be imagined.'[25] Between gales, the men built two huts on the pebbly beach to accommodate the ten men who would remain as the first humans to winter on the continent. Their living quarters were insulated against the cold with a double floor and walls lined with sheets of papier-mâché, and a loft insulated with seal skins. Two tiny rooms lined with fur and wool became a darkroom for developing photographs and a space for undertaking taxidermy.[26] Within weeks of the departure of the *Southern Cross*, the bay filled with ice and the penguins were rapidly departing for their winter feeding grounds in warmer waters.

As the polar winter descended, these first winterers began to feel the psychological stresses associated with constant darkness, isolation and confinement. Bernacchi found himself alternating between feeling spellbound and horrified by his predicament. He confided to his diary that men and dogs alike were becoming increasingly irritable and morose, while Borchgrevink, who showed himself to be a meddling and impractical leader, was generally loathed by his men and becoming increasingly isolated from the group.[27]

← Tasmanian physicist and astronomer Louis Bernacchi stands at the entrance to one of the *Discovery's* two magnetic huts during the British National Antarctic Expedition in 1904. Understanding the nature of terrestrial magnetism in the region was a key part of the expedition, which contributed to synchronised observations with Erich von Drygalski's German *Gauss* expedition and Dr Otto Nordenskjold's Swedish *Antarctic* expedition.

← Louis Bernacchi at Cape Adare in 1899. He was the first Australian to winter in Antarctica.

Finally, as midwinter approached, Borchgrevink's bizarre behaviour pushed Bernacchi to breaking point. 'At last matters have reached a climax,' he wrote. 'My opinion that Borchgrevink is insane now confirmed. Really believe him to be insane.'[28] His entry for Christmas 1899 revealed his desperation: '(Xmas Day—in Purgatory.) All over the world a time of rejoicing … Here misery, desolation and a flood of nostalgia to drive one mad.'[29]

By the time the *Southern Cross* returned to pick up the group in January 1900, the party had been on the ice for 11 months. Despite the prolonged anxiety and hardships they had endured, the men showed that it was possible for humans to survive an Antarctic winter. Their scientific achievements were also notable: completing a sledging journey to the highest southern latitude of any previous expedition, at latitude 78°50' S; locating the position of the south magnetic pole; collecting the first samples of insects and seaweed from the frozen continent for the Natural History Museum in London; and recording Antarctic weather and magnetic data over a whole year.[30] The zoologist Nicolai Hanson, who tragically fell ill and died in the hut, had made extensive observations and gathered numerous marine specimens during the expedition, but his zoological records went missing and were never recovered. Remarkably, Bernacchi's own careful meteorological and magnetism observations survived, offering tantalising insights into the physical nature of the southern polar region.

The 'Daintree' of Antarctica

Vegetation is a rare sight in Antarctica. Less than one per cent of the continent is free of ice, so visitors to Australia's Casey station on the coast of East Antarctica would be surprised to see patches of verdant green among the stark white ice and snow. The summers here are short and all life must tolerate extreme winds and freezing temperatures to survive. The only plants that can grow in these conditions are mosses, liverworts, lichens and fungi. Casey has one of the rarest ecosystems on the planet. Here, lush, slow-growing moss beds—some more than 100 years old—thrive in the coastal environment, drawing on nutrients left where penguins once congregated in their thousands. This has earned it the unofficial title of the 'Daintree' of Antarctica, a reference to the densely vegetated Daintree rainforest in Queensland's Wet Tropics.[31] These mosses are the largest plants on the continent. They are Antarctica's version of old growth forests and, like tree rings, they harbour a secret. Within their shoots there is a unique archive of past climatic conditions on the polar continent. In recent years, Australian scientists have been monitoring the health of the Casey moss beds, and the news is not good. One study conducted in 2018 found that, over just 13 years, these moisture-loving plants were showing significant signs of stress, as colder summers and stronger winds brought drier conditions. These ancient moss beds have become a sentinel species for the effects of climate change in East Antarctica.[32]

↓ A rare patch of green among the ice: the moss beds near Casey station are ecologically significant and highly vulnerable to changing climatic conditions.

The question of Antarctica

When the *Southern Cross* returned to Hobart on 16 April 1900, the expeditioners were greeted by cheering crowds and telegrams of congratulations from around the world. Australia was preparing for Federation and, by the time the first Australian Parliament was sworn in on 9 May 1901, the mysterious frozen continent to the south was firmly in its sights.[33] One of the first expeditions to arouse Australia's Antarctic intrigue was Ernest Shackleton's *Nimrod* expedition to the Ross Sea in 1907–1909, which several Australians were invited to join. Among them was Captain John King Davis, a British-born Australian seaman appointed as chief officer aboard the steam yacht, and a Welsh–Australian geologist, Professor (Sir) Tannatt William Edgeworth David, who successfully appealed to the Australian Prime Minister Alfred Deakin for financial support for the expedition.[34] David had argued:

The great southern continent, known as Antarctica, should be of more interest and importance to Australia than to any other country on account of its being our nearest neighbour, and on account of its control of our Australian weather conditions, and of the magnetic conditions which govern navigation in our southern seas.[35]

David, known to his colleagues as 'The Professor', selected two of his former geology students to join him: Douglas Mawson, who had a keen interest in South Australia's glacial geology, and Leo Arthur Cotton. So began Mawson's long and influential association with Australian Antarctic exploration and science.[36]

← The Welsh-Australian geologist and Antarctic explorer Tannatt William Edgeworth David during the *Nimrod* expedition. Despite being much older than his companions, 'The Professor' led the first ascent of Mount Erebus, Antarctica's only active volcano, and later made an epic four-month sledging journey to the south magnetic pole with Douglas Mawson and Alistair Mackay.

↑ The *Nimrod* docking next to sea ice during the British Antarctic Expedition led by Ernest
Shackleton between 1907 and 1909. The expedition wintered on Ross Island in McMurdo Sound
after ice prevented the ship from reaching the intended base on Edward VII Peninsula.

On arrival at Cape Royds, David prepared to climb Mount Erebus, the southernmost active volcano on Earth. He chose Mawson and Alistair Mackay, a young Scot who had served as a surgeon in the Royal Navy, to accompany him. In spite of their collective lack of climbing experience and the onset of a blizzard during their ascent, the three managed to scale the ice ramparts of Mount Erebus and examine the inside of the crater.[37] Later, after spending winter in the expedition hut, David and his two companions set out to reach the south magnetic pole somewhere to the north-west of Cape Royds. The south magnetic pole is a point where Earth's magnetic field is vertical. It is not a fixed location,

↓ Members of the expedition and crew standing on sea ice in front of the docked *Nimrod* at Cape Royds landing site, Ross Island, Antarctica, 1908.

← David's 1908 photograph of Mount Erebus erupting, with the Ross Island hut in the foreground.

↓ The dip circle was used to measure the angle between the horizon and Earth's magnetic field (the dip angle). The dip needle pointed directly downward when positioned directly over a magnetic pole.

but rather moves in response to Earth's magnetic field. For David and Mawson—both geologists—the quest to reach the elusive south magnetic pole was tantalising. James Clark Ross had managed to reach the north magnetic pole in 1831, but was thwarted ten years later when he attempted to locate the south magnetic pole, his passage blocked by the Great Ice Barrier (now known as the Ross Ice Shelf).

David's party left the hut on Ross Island on 5 October 1908, planning to return by 15 January 1909. They spent a week crossing the frozen sea ice of McMurdo Sound, hauling sledges and experimenting with makeshift sails, before negotiating the fragile sea ice along the coastal margin of Victoria Land. By 30 October, they had barely completed one-third of the distance to the south magnetic pole. Sastrugi (ridges of wind-hardened snow) sabotaged their sledges, and they were forced to supplement meagre rations by killing seals and penguins. David, almost twice the age of the other two, found the journey exhausting, and tensions escalated between him and Mawson. On 1 December, David described their progress:

> *For half a day we struggled over high sastrugi,*
> *hummocky ice ridges, steep undulations of bare blue*
> *ice with frequent chasms impassable for a sledge,*
> *unless it was unloaded and lowered by Alpine rope. After*
> *struggling on for a little over half a mile we decided to*
> *camp, and while Mawson took magnetic observations*
> *and theodolite angles, Mackay and I reconnoitred ahead*
> *for between two and three miles to see if there was any*

way at all practicable for the sledge out of these mazes of chasms, undulations and séracs [sharp ridges]. Mackay and I were roped together for this exploratory work, and fell into about a score of crevasses before we returned to camp, though in this case we never actually fell with our head and shoulders below the lids of the crevasses, as they were mostly filled at the surface with tough snow. We had left a black signal flag on top of a conspicuous ice mound as a guide to us as to the whereabouts of the camp, and we found this a welcome beacon when we started to return, as it was by no means an easy task finding one's way across this storm-tossed ice sea, even when one was only a mile or two from the camp.[38]

During the four-month journey, they traversed 2,028 kilometres, manhauling their sledges across the sea ice and up the Drygalski Ice Tongue and Larsen Glacier to the polar plateau, more than 2,140 metres above sea level. On 16 January 1909, at latitude 72°25' S, longitude 155°16' E, they were at the location Mawson calculated for the south magnetic pole. Here the exhausted men planted a Union Jack in the snow, and David declared the surrounding land to be part of the British Empire.

Returning to their depot, the men celebrated their achievement with a serving of 'hoosh' (a thick stew made with biscuits and a mixture of dried and ground meat and fat, also known as pemmican) before beginning the return journey to rendezvous with the *Nimrod*. 'After so many days of toil, hardship and danger,' David wrote, '… we were too utterly weary to be capable of any exaltation.'[39] The return journey presented its own challenges. David's health deteriorated, so Mawson assumed the role of leader. The party reached the coast after 122 days, having completed the longest unsupported manhauling sledge journey attempted in Antarctica. Along the way, Mawson continued his scientific work, charting the coastline and making magnetic and geological observations. David would later pay tribute to Mawson as 'the real leader who was the soul of our expedition to the Magnetic Pole. We really have in him an Australian Nansen, of infinite resource, splendid physique, astonishing indifference to frost.'[40]

→ Douglas Mawson about to descend a crevasse at Cape Royds, Ross Island, during the *Nimrod* expedition in 1908.

We had found an accursed country. On the fringe of an unspanned continent along whose gelid coast our comrades had made their home—we knew not where—we dwelt where the chill breath of a vast, Polar wilderness, quickening to the rushing might of eternal blizzards, surged to the northern seas.

SIR DOUGLAS MAWSON[1]

Coast

← Pancake ice at Petersen
Bank, Antarctica.

Douglas Mawson's fascination with Antarctica remained undiminished and, in December 1911, he returned to the continent as leader of the Australasian Antarctic Expedition (AAE) aboard the *Aurora*, a 35-year-old sealing vessel. During his campaign to obtain government funding for the expedition, Mawson stressed the strategic importance of the region to Australia as a potential source of wealth.[2] He also saw it as an opportunity for Australia to demonstrate its worth as a new nation.

> *For many reasons ... I was desirous that the Expedition should be maintained by Australia. It seemed to me that here was an opportunity to prove that the young men of a young country could rise to those traditions which have made the history of British Polar exploration one of triumphant endeavour as well as of tragic sacrifice. And so I was privileged to rally the 'sons of the younger son'.*[3]

Mawson's main objective was to establish three scientific bases—one on Macquarie Island, halfway between New Zealand and the Antarctic continent, that would also serve as a radio relay station, and two on the continent itself—the Main Base in Adélie Land under Mawson's leadership, and the Western Base in Queen Mary Land under the leadership of British seaman John Robert Francis (Frank) Wild, who had already established his leadership credentials during Shackleton's Antarctic expeditions.[4] At each base the expeditioners would undertake scientific research on geology, cartography, meteorology, geomagnetism and biology, while Captain John King Davis, the Australian master of the *Aurora* and second-in-command of the expedition, oversaw a program of ocean studies from the deck of the ship. Captain Davis had previously served as first officer

aboard the *Nimrod* during Shackleton's Antarctic expedition in 1907–1909. He later served as commander of the Ross Sea Relief Expedition to rescue Shackleton's 'shore party' from McMurdo

← John King Davis stored records in this trunk from his Antarctic voyages. Australia's second continental station was named Davis in recognition of his significant role in Australia's early expeditions in the Antarctic region.

↑ Crowds gathered on the Hobart wharf in 1911 to farewell members of the AAE aboard the *Aurora*, bound for the little known southern polar region.

Sound in 1916–1917, and as commander of the *Discovery* during the first voyage of Mawson's British, Australian, New Zealand Antarctic Research Expeditions, known as BANZARE, in 1929–1930. He was awarded the King's Polar Medal for his services to polar exploration.[5] Australia's second Antarctic continental station, established in 1957, was named in his honour.

Adélie Land

Many of the expeditioners began writing diaries as they embarked on their voyage south, capturing their impressions of the Antarctic environment and tantalising details of daily life.[6] Four of the men were from Sydney, including Charles Francis Laseron, a geologist with the Technological Museum who was appointed to the Main Base party as a taxidermist and general collector and participated in two sledging journeys between 8 November 1912 and 6 January 1913. Like many other explorers of the era, Laseron suffered from severe seasickness during the voyage, describing the first leg to Macquarie Island as a 'horrible nightmare'.[7] By the time the ship approached the coast and dropped anchor in Commonwealth Bay on 8 January 1912, his disposition had improved. He observed 'an excellent little harbour' with shallow water and a series of small islands topped with snow like 'tasty Christmas cakes'.[8] Later he recalled the moment of arrival:

> On either side of the peninsula, now called Cape Denison, the shores of Commonwealth Bay stretched in a great semicircle, bordered everywhere by high ice cliffs, with here and there a patch of black rock showing at the base. Inland a steep slope led upwards to the plateau directly behind the hut, and up this slope sledging parties would ultimately find a way, to explore the new lands, east, west and south.[9]

Captain Davis was pleased to have found an ice-free site for the Main Base Hut or 'Winter Quarters'. Such ice-free rocky areas comprise just five per cent of the continent's vast coastline; the remainder of the coast is covered with ice shelves and ice cliffs. 'Here was the first locality we had seen in our coastwise search where the red and brown bones of the last continent appeared, free of their frozen blanket,' wrote Davis.[10] As the expeditioners set about constructing the Main Base, however, they were soon engulfed by frequent blizzards and ferocious katabatic winds, formed by cold, dense air rushing down from the interior plateau to the steep cliffs along the coast. Indeed, wind would become the defining feature of the expedition, as the Australian meteorologist Cecil Thomas Madigan observed:

> The wind was the most remarkable feature of the meteorology—or indeed of the locality. It is the outstanding characteristic of Adélie Land. Commonwealth Bay is

↑ While Mawson's men finished the Main Base Hut in 1912, they created a temporary camp known as 'Benzine Hut' from stacked wooden cases used for transporting cans of motor fuel to Antarctica.

probably the windiest place on the earth, and certainly it appears to be so as far as records up to the present indicate. For nine months of the year an almost continuous blizzard rages, and for weeks on end one can only crawl about outside the shelter of the hut, unable to see an arm's length owing to the blinding drifting snow.[11]

The art of hurricane walking

Cecil Madigan had the unenviable job of recording weather conditions every six hours in these terrible conditions. He would peer at the barometer and thermometer screens in order to record wind direction and velocity, as well as observing cloud behaviour, snow, sea conditions and the astonishing displays of the aurora australis.[12] In his meteorological day book he described how: '[a]fter some practice the members of the expedition were able to abandon crawling, and walked on their feet in these 90 mile [145 kilometres per hour] torrents of air, "leaning on the wind".[13] Mawson concurred, calling it the art of 'walking in hurricanes':

It may well be imagined that none of us went out on these occasions for the pleasure of it. The scientific work required all too frequent journeys to the instruments at a distance from the Hut, and, in addition, supplies of ice and stores had to be brought in, while the dogs needed constant attention ... Whatever has been said relative to the wind-pressure exerted on inanimate objects, the same applied, with even more point, to our persons; so that progression in a hurricane became a fine art. The first difficulty to be encountered was a smooth, slippery surface offering no grip for the feet. Stepping out of the shelter of the Hut, one

was apt to be immediately hurled at full length down wind. No amount of exertion was of any avail unless a firm foothold had been secured. The strongest man, stepping on to ice or hard snow in plain leather or fur boots, would start sliding away with gradually increasing velocity; in the space of a few seconds, or earlier, exchanging the vertical for the horizontal position. He would then either stop suddenly against a jutting point of ice, or glide along for the twenty or thirty yards [18–27 metres] till he reached a patch of rocks or some rough sastrugi ... Ensconced in the lee of a substantial break-wind, one could leisurely observe the unnatural appearance of others walking about, apparently in imminent peril of falling on their faces.[14]

The weather continued to deteriorate as winter approached. When the penguins left for their winter feeding grounds in the Southern Ocean, the landscape became even more desolate. The frequency of snowfalls soon buried the hut, and the men found it necessary to dig a series of tunnels to gain access. In finer weather, they could exit via a trapdoor in the roof of the verandah, but getting in and out of the hut during a blizzard required crawling through a low opening. Eventually snow smothered everything, making it difficult for those working outside to even find the entrance. As Mawson recalled, '[a] journey by night to the magnetic huts was an outing with a spice of adventure':

Climbing out of the veranda, one was immediately swallowed in the chaos of hurtling drift, the darkness sinister and menacing ... Unseen wizard hands

↑ The Main Base Hut at Cape Denison comprised a living hut for sleeping, kitchen, dining, laundry, storage and darkroom facilities for 18 men, with a workshop leading off it.

↖ 'Commonwealth Bay is probably the windiest place on Earth.' So wrote meteorologist Cecil Madigan during the AAE. Frank Hurley's photograph captured it perfectly, showing a fellow expeditioner 'leaning on the wind' as he picked ice for 'culinary purposes'.

clutched with insane fury, hacked and harried … Cowering blindly, pushing fiercely through the turmoil, one strove to keep a course to reach the rocks in which the huts were hidden—such and such a bearing on the wind—so far. When the rocks came in sight, the position of the final destination was only deduced by recognising a few surrounding objects.[15]

Each man would take his turn going outside to tend to the dogs or to make scientific observations. As the drifts became thicker, it was possible to become disoriented and be standing on the roof of the Main Base Hut without realising it, until accidentally falling through it into the verandah. Mawson wrote:

Picture drift so dense that daylight comes through dully, though, maybe, the sun shines in a cloudless sky; the drift is hurled, screaming through space at a hundred miles an hour [160 kilometres per hour], and the temperature is below zero, Fahrenheit [-18°C]. You have then the bare, rough facts concerning the worst blizzards of Adélie Land. The actual experience of them is another thing … Shroud the infuriated elements in the darkness of a polar night, and the blizzard is presented in a severer aspect. A plunge into the writhing storm-whirl stamps upon the senses an indelible and awful impression seldom equalled in the whole gamut of natural experience.[16]

Laseron's work required him to record the physical geology of the area around the Main Base, collect lichens and mosses, and help to dredge biological specimens from the water in the bay. On Friday 22 March 1912, he recorded that the winds had not blown less than 60 miles per hour [97 kilometres per hour] for a whole month. With sudden gusts of up to 100 miles per hour [160 kilometres per hour] and whirlwinds from every direction, they could knock a man over, even as another was standing nearby in perfectly calm conditions. Several days later, the cartographer and meteorological assistant Alfred J. Hodgeman became lost while out reading the anemometer, and spent all morning trying to find his way back to the Hut. By May, the wind velocity had increased even further, making scientific observations and sampling almost impossible. Laseron concluded that the continent was 'the most desolate, cruellest region in the world'.[17]

↑ Mawson's men dug tunnels to access the outside world from the Main Base Hut. Here an expeditioner, returning from his duty as nightwatchman, pushes his way into the verandah through rapidly accumulating drift snow.

The Western Party

Meanwhile, the *Aurora* had sailed westward from Cape Denison in search of a suitable site for the Western Base. It was already late in the season as the ship steamed along the coastline, buffeted by frequent blizzards and slowed by heavy pack ice. Time was running out when Frank Wild settled on a site, 585 metres from the edge of an immense sheet of ice that stretched 320 kilometres east and west along the coast, jutting some 20 kilometres into the sea. This prominent feature of East Antarctica was first sighted by the United States Exploring Expedition in 1840 and named Termination Land. Mawson called it the Shackleton Ice Shelf. The group of eight men—all Australian apart from Wild—would spend the winter on floating ice, 27 kilometres from land. Wild wrote in his diary:

> *Some of the ships company gave their opinions to the Australian press and at least one paper stated that 'Wild's party is camped on moving ice & there is little probability they will ever be seen again'.*[18]

↓ The Western Party completed the hut they called 'The Grottoes' within a week, but it was soon engulfed by blizzards and deep snow piled up around the hut so that only the peak of the roof was visible.

↓ Frank Wild described the winter routine for his Western Party, with every member having a job to do: 'Work commenced at 10 a.m. during the winter, & finished at 1 p.m. unless anything special had to be done. Divine Service was held every Sunday at which Moyes & I officiated in turn'.[19]

Nevertheless, Mawson had been careful in selecting the members of the Western Party, aware that they would have to face a year in extreme conditions and isolation. Under Frank Wild's leadership, the group comprised George Harris Sarjeant Dovers (cartographer), Charles Turnbull Harrisson (biologist and artist), Charles Archibald (Arch) Hoadley and Andrew Dougal Watson (geologists), Dr Sydney Evan Jones (medical officer), Alexander Lorimer Kennedy (magnetician) and Morton Henry Moyes (meteorologist). Mawson described that first day of the AAE's Western Base:

> As the Aurora receded in the distance; the men of Wild's party ceased gazing hypnotically after her and turned their faces cheerfully and hopefully towards the shelf-ice where they were to be domiciled for good or ill for at least a 12-month period before any relief could arrive. On that morning, the 2-first of February, 1912, with their private luggage and some essential articles, they sledged from the brink of the ice precipice across some 640 yards [585 metres] to the site, selected as reasonably free from small crevasses, whereon the Hut was to be built. The job of erecting the structure was then immediately undertaken. The weather remained good but the temperature fell to 5°F [-15°C] during the night.[20]

← A sketch of the Western Base hut by the Tasmanian-born biologist Charles Harrisson.

Wintering

As the dark winter months descended, the men were kept busy—weather permitting—with their allotted duties. The Western Party spent the first week building a hut that would shelter them as they wintered on the Shackleton Ice Shelf. They called it 'The Grottoes'. Wild wrote:

> I now thought it time to establish a winter routine. Each member had his particular duties to perform, in addition to general work, in which all hands were engaged. Harrisson took charge of the lamps and checked consumption of oil. Hoadley had the care of the provisions, making out lists showing the amount the cook might use of each article of food, besides opening cases and stowing a good assortment on convenient shelves in the veranda. Jones and Kennedy worked the acetylene plant ... Jones, in addition to his ability as a surgeon, showed himself to be an excellent plumber, brazier and tinsmith, and the Hut was well lighted all the time we occupied it. Moyes's duties as meteorologist took him out at all hours. Watson looked after the dogs, while Dovers relieved other members when they were cooks.[21]

At the Main Base, Mawson also established a routine of sorts—amidst the general chaos that prevailed in the dimly lit living hut in which the men spent much of their time between blizzards:

> Our hearth and home was the living Hut and its focus was the stove. Kitchen and stove were indissolubly linked, and beyond their pale was a wilderness of hanging clothes, boots, finnesko [reindeer-skin boots], mitts and what not, bounded by tiers of bunks and blankets, more hanging clothes and dim photographs between the frost-rimmed cracks of the wooden walls.[22]

Their days were filled with noise: from the rousing wakeup call by the nightwatchman, gramophone music and the clatter of cooking pans, to the incessant sounds of wind and creaking hut timbers. The men grew so accustomed to it that they were uneasy if things became too quiet. The atmosphere in the living hut was a kind of wilderness in its own right; although, as Mawson put it, 'raucous good humour prevailed over everything'.[23]

↑ An interior view of the kitchen area in the Main Base Hut at Cape Denison.

↑ During his year at Cape Denison, photographer Frank Hurley documented daily life and routines in the Main Base Hut. Here the astronomer and assistant magnetician, Robert Bage, mends his sleeping bag in a corner of the living hut.

← A container labelled 'compressed mixed vegetables' produced for Antarctic expeditions by Swallow & Ariell Limited, Port Melbourne, Australia.

↘ Special occasions, such as birthdays and the traditional midwinter feast shown here, were important for maintaining morale and enhancing group cohesion.

Cooking mishaps were a source of particular amusement, with those responsible bestowed titles such as 'Assistant Grand Past Master of the Crook Cooks' Association'. Cooking, under the inspiration of Mrs Beeton, became a fine art, and special occasions such as birthdays called for the cleanest clothes and a sumptuous menu, toasts and extravagant speeches.[24] Mawson observed:

> The mania for celebration became so great that reference was frequently made to the almanac. During one featureless interval, the anniversary of the First Lighting of London by Gas was observed with extraordinary éclat.[25]

The expedition's chief medical officer and bacteriologist, Dr Archibald Lang (Archie) McLean, was interested in the men's physical response to such extreme conditions, conducting monthly checks of six of them to monitor changes in their blood. Outside activities were necessarily more prosaic. There was ice-carrying, dog-tending, snow-shovelling, coal-carrying, retrieving stores, constructing the huts for recording weather and magnetism, digging ice shafts to study the sea ice, glaciers and lakes, and collecting seals and birds as food and zoological specimens. The men invented their own instrument called a 'puffometer' to measure the speed of strong wind gusts, hauling it to the top of 'Annie Hill' near the Main Base using a pulley and line. As Mawson recalled:

The puffometer was left out for an hour at a time, and separate gusts up to one hundred and fifty and one hundred and eighty miles per hour [240–290 kilometres per hour] were commonly indicated. I remember the final fate of this invention. While helping to mount it one day, the wind picked me up clear of the ground and dashed myself and the instrument on some rocks several yards away. The latter was badly damaged, but thick clothing saved me from serious injury.[26]

Between blizzards, the men at Main Base tried to erect wireless masts with the intention of establishing a radio link between Antarctica and Australia via a relay station on Macquarie Island. The extreme conditions at Commonwealth Bay hampered their efforts for nearly a year until, finally, in February 1913 they succeeded in making the first radio transmission from Antarctica.

Polar pastimes

Antarctica's extreme isolation and harsh conditions can take a mental toll. Indeed, maintaining morale and group cohesion is as important as physical safety.[27] For Mawson's men, confined to their two-roomed hut at Commonwealth Bay over long winter months, leisure activities helped to ease worried minds. Charles Laseron noted in his diary: 'taking it all in all, time is passing pleasantly enough, during the day we are all busy, and after the evening meal, we are free to read or write or go to bed, the last being very acceptable'.[28] In order to alleviate anxieties associated with boredom and maintain morale, whether confined to small huts or sledging across the vast inland plateau, the men would spend the long hours of darkness writing in their diaries, listening to gramophone records, or creating their own plays, short stories, songs and poetry satirising their current predicament or retelling a recent adventure.

In October 1912, the men at Main Base held an impromptu drama to celebrate Xavier Mertz's birthday. They enjoyed it so much, they decided to stage a 'Grand Opera', *The Washerwoman's Secret*. According to its author, Laseron, it was 'a tragedy in five acts, with a complicated and highly dramatic plot' and, he reported, a 'howling success ... we have managed to extract some fun from this God forsaken hole'.[29]

Reading was a popular pastime. The men had brought an assortment of books, from novels and poems to technical manuals, newspapers and accounts written by earlier explorers, which they often read aloud and discussed. Cookery books were of particular interest, especially when food was scarce, and Mrs Beeton's *Book of Household Management* was a favourite.[30] Books were also an essential item to pack for sledging journeys, when the travellers might be confined to a small tent for days on end during a blizzard. Music played its part too. Some men had brought instruments with them—a folding organ, concertina, flute, piccolo and mouth organ. They also listened to gramophone records and sang hymns and popular songs. Their 'Adélie Land Band' was so popular that the men crawled out of their bunks to be part of it.

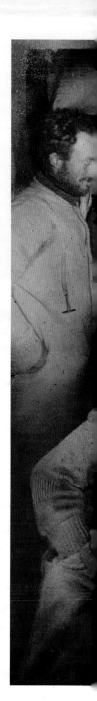

↓ A winter evening at the Main Base Hut during the first year of the AAE.

This wild, weird dead land

One of the main objectives of the AAE was to undertake a series of sledging journeys before the *Aurora* returned in January 1913. The sledging parties from their respective bases on the Shackleton Ice Shelf and Commonwealth Bay would explore and chart the coastline, glaciers, ice shelves and sea ice, as well as record observations of the environment and wildlife along the way. With the arrival of the penguins signalling the start of spring and the promise of better conditions, support parties set out to investigate the surrounding terrain and establish food depots in preparation for the main traverses in November. These early forays also gave the men an opportunity to gain some sledging experience, and to learn first-hand about the capricious conditions out on the ice sheet. Tasmanian-born biologist and artist Charles Turnbull Harrisson, appointed to Wild's Western Party, was relieved to be starting his first sledging adventure after the pall of the winter months. 'Glad to get a go on for the sake of warmth,' he confided in his diary. 'It is terrible on the fingers and feet.' Within a few hours, however, the plateau began to reveal its true nature as the party encountered its first field of crevasses, about a mile long. Harrisson wrote:

> *Walls of ice—its depth we could not see for snow lodged about 50 ft [15 metres] down. A depth of blue. The ends, hung with icicles—and beneath the lodged snow, a glimpse of walls coming together in a lower depth—that might have led down to Dante's lowest hell of cold!*[31]

Having safely navigated the crevasses, the men found themselves hauling their sledges through the wave-like ridges of sastrugi. It was hard going, and Harrisson found the plateau tiresome: 'The same "stark and solemn

← Andrew Watson captured the moment as his Western Base sledging party guided a heavy-laden dog sledge over a gully.

solitude" as of all this wild weird dead land', he confided to his diary. A beautiful sunset gave him some pleasure, but his party's immediate concerns were far less ethereal: damp clothes from sweat, frozen air in the tent that condensed and dripped onto clothes and reindeer bags. 'Nothing dries here now', Harrisson complained. His feet also ached from the cold and the walking.[32] A few days later, Wild observed an ominous line of cloud along the horizon and ordered everything to be stowed in the tents, and holes dug in the snow for the dogs. The men sheltered in the tents as the blizzard struck. Occasionally, someone would venture outside to feed the dogs or push away snowdrift that had built up around the tent walls. On 1 September, Harrisson recorded his displeasure of sledging in 'flimsy tents', drawing a wry comparison between the Antarctic spring and the Australian spring of his childhood:

> Spring! 'September with fair yellow tresses' if I remember the line right—in this dead land no golden wattle marks the returning spring and September's yellow tresses are the blizzard's cirrus streaming across the sun![33]

The storm abated after a week of confinement, and the sledging party resumed its journey. The Antarctic winds, however, seemed oblivious to the coming of spring. On one occasion, a gust tore the tent and a stove blew away. Harrisson recovered the stove 275 metres away, together with other gear that had been scattered by the wind. Without the tent, the men dug a hole in the ice for shelter, and Harrisson spent the day there reading *The Letters of Robert Louis Stevenson*, relieved to not have to listen to the constant sound of flapping canvas.[34] Nevertheless, avalanches of ice crashed down from nearby cliffs, and everyone's nerves were on edge. Such extremes of Antarctic life were not lost on Harrisson:

> Toil and laziness, suffering and comfort! Privations, bitter cold, hard dragging until you feel done up. Whole days of idleness, sleeping the hours away [then] long stretches of unvarying monotony; beautiful, weird—sometimes majestic—scenery.[35]

The party returned safely to their hut, but Harrisson was soon off on another sledging journey, this time to establish a food depot for a longer traverse to be led by Wild. It was a bright, beautiful morning on 30 October as the two teams—the 'eastern party'

Through the western fringe of the Bergshrund Pressure

Dec 7th 1912

(Antelopes on S. side of Bells island)   (Cliffs rock)

↑ 'Through the western fringe of the Bergshrund Pressure' (a glacial crevasse)
sketched by Charles Harrisson, a member of the Western Party and a keen artist
who once sketched until his fingers were frostbitten.

comprising Wild, Watson and Kennedy, and the 'supporting party' comprising Harrisson
and three dogs—posed for photographs after a breakfast of seal's liver, bacon and gravy.
Then they were off to the sounds of 'God be with you till we meet again' wafting from
the gramophone, leaving Morton Moyes alone at the hut. Harrisson wrote in his diary:

> *Out on the White Waste—the limitless line of the Glacier before us. Wild on his long
> journey, hoping to do 500 miles [800 kilometres] before turning back. I hoping to
> accompany them for the first 100 [160 kilometres]. Left hut about 10.30 a.m. Would
> have been a hard pull, but for the supporting party & the dogs—for the 'Eucalypt'
> [his Australian-made sledge], in addition to my food & baggage (& dog's biscuits)
> is carrying 165 lbs [75 kilograms] of the other party's food. Made good progress.
> Wild in lead. I with dogs.*[36]

It was intended that Harrisson, sledging solo, would return to base to rejoin Moyes after
establishing a food depot. When he lost his sledge in an accident, however, Harrisson had

no choice but to continue on with Wild's party and their remaining sledge. Meanwhile, Moyes spent ten weeks in a lonely vigil, awaiting Harrisson's return. Moyes's diary reveals his mounting anxiety:

> *Monday Nov 4th*
>
> *Blew very heavy all night & a heavy gale all day. Went out twice, but found it difficult to stand against the wind, & drift stung my face badly. A quiet day indoors. Not much cooking, & all reading. They should have left to-day. George gave me a packet of letters to give his Father if he has an accident, but I hope to return them to him in 3 months. Had a glorious bath tonight, first for about 6 weeks. Snow got thro' the window & roof last night. Found about 1 ft [30 centimetres] deep on the table when I got up. Reading Dante's Inferno. Temp +20. Bar 29.80.*
>
> *Tuesday Nov 5th*
>
> *Light wind & falling snow. Out to the Thermograph & heaved in tin of Glaxo. Read Dante's Inferno all day. Got 'em Stuffed Seal's Heart for dinner. Still light outside at 9.30 p.m.*[37]

Summer sledging

By early November, despite continuing poor weather at Commonwealth Bay, Mawson too was anxious to begin the main sledging journeys before the *Aurora* arrived in mid-January 1913. His plans for summer sledging included five expeditions. A Southern Party, comprising astronomer and engineer Robert Bage as leader, the photographer Frank Hurley and New Zealand engineer Eric Webb, would focus on magnetic observations in the vicinity of the south magnetic pole, supported by Herbert Dyce Murphy (leader), John George Hunter and Laseron, who would accompany the party as far as possible and return to the Main Base by the end of November. A Western Party, led by the engineer Francis Bickerton, would traverse the coastal highlands west of the Main Base using an air-tractor sledge that Bickerton had fashioned from the body of the expedition's fixed-wing aircraft, which had crashed during a test flight.[38] The Near East Coast Survey Party, with Frank Stillwell (leader and geologist), John Close and Hodgeman, would map the

↑ The Southern Supporting Party photographed by Frank Hurley at the Southern Cross Depot (from left: John Hunter, Herbert Murphy and Charles Laseron).

coastline between Cape Denison and the tongue of the glacier later named for Dr Xavier Mertz, and support the other eastern parties working further afield. An Eastern Coastal Party, comprising Madigan (leader), McLean and Percy Correll, would investigate the coastline beyond the Mertz Glacier. And a Far Eastern Party, led by Mawson himself, would take dogs and set off overland to the south of Madigan's party, mapping more distant sections of the coastline. He would take Belgrave Ninnis and Mertz, 'both of whom had so ably acquitted themselves throughout the Expedition and, moreover, had always been in charge of the dogs'.[39] Mawson instructed each party to be back at the Main Base by 15 January 1913 to rendezvous with the *Aurora*, and noted:

Geographical investigations by means of sledge journeys are regarded as the best form of Antarctic scientific research and you are encouraged to push this work as far as possible. On sledging journeys complete accounts of all natural phenomena should be kept. It will be found advisable to take the ruled sheets and fill in your discoveries and tracks as you proceed. The delineation of coastline should occupy a first place in your sledging program. A short journey onto the plateau will lead to valuable results. A party should attempt delineation of coastline westward and another eastward of the base.[40]

← Two members of the Western Party carrying out 'experimental work' on the ice sheet.

Each of the summer sledging parties had their fair share of blizzard conditions and dramatic incidents. The Southern Party, comprising Hurley, Bage and Webb, set off from the Main Base for the south magnetic pole on 10 November 1912, manhauling their heavily loaded sledge for nine weeks across the unexplored plateau. Much of the outward journey was uphill, with frequent blizzards, although Webb managed to undertake a complete series of magnetic observations. On 16 December, the weather suddenly changed for the better. Hurley wrote:

> It was hard to imagine we were 250 miles [400 kilometres] on the plateau with nearly 250 miles of ice separating us from the hut. There was not even enough wind to stir the tent & although zero, was warm. I thought it seemed as if camped in the Australian bush & was only brought back to Antarctica by the vigorous boiling of the cooker. At lunch we were all merry & sang old favourite ditties. What a contrast to our usual conditions.[41]

The weather deteriorated on the return journey, and they missed the 67 mile [108 kilometre] depot altogether, leaving them perilously short of food and having to make a dash back to the hut at Cape Denison. They arrived just as the *Aurora* was sailing into Commonwealth Bay.

Murphy's party, which returned to Main Base on the same day as Stillwell's, had endured severe snowdrift and all three men were suffering from snow blindness. Mawson made some adjustments to his sledging plans, assigning Hodgeman to the Western Party with Bickerton and Dr Leslie Whetter who would travel with the modified air-tractor towing a train of four sledges. Bickerton had already made several successful test runs with the mechanised sledge, and the party set out in high spirits on 3 December 'amid an inspiriting demonstration of goodwill from the six other men' who remained at the Hut.[42] Within a short time, however, the engine began misfiring and progress became excruciatingly slow. Bickerton made efforts to keep it moving but, on the second day out from the Main Base, the engine jerked to a halt and the propeller disintegrated. The air-tractor sledge was abandoned and the three men continued on, manhauling their sledges. It was on this traverse that Bickerton discovered a small black rock lying in a shallow depression about 32 kilometres west of the Main Base. It was the first meteorite to be found in Antarctica.[43]

On 9 December, Stillwell, Laseron and Close set out for another foray, this time along the coast immediately east of Commonwealth Bay; however, the conditions soon deteriorated and they were forced to seek shelter in Aladdin's Cave, a cavern dug into the ice by Mawson, Ninnis and Madigan during the previous winter to serve as a depot and refuge for the sledging parties. Sheltering deep in the ice, however, came with its own risk. Aware of the dangers of poor ventilation, the men tried to keep the cave entrance clear, but by morning they were overcome with fumes. When Stillwell, cooking hoosh on the primus stove, collapsed, Close thrust an ice-axe into the choked entrance before he too collapsed. Laseron was too weak to stir. The small hole made by Close's axe was enough to save them, and the three men managed to crawl out and recover sufficiently to complete their survey of the coast and islets east of Commonwealth Bay.[44]

By 17 January 1913, all the summer sledging parties had returned except for Mawson's Far Eastern Party, which had left on 10 November 1912. The party of three had taken two dog teams with the intention of exploring the coastal region east of Commonwealth Bay, and they were well equipped with sufficient rations and food depots for their two-month journey. Their route would take them across two heavily crevassed glaciers and, despite several falls into crevasses, they managed to cover over 500 kilometres in their first few weeks. By 14 December, the men were in good spirits and planned to make a short dash eastward before turning back. In the early hours of the morning, tragedy struck. Ninnis, walking next to his sledge and dog team, disappeared into a crevasse. Mawson later wrote:

> I leaned over and shouted into the dark depths below. No sound came back but the moaning of a dog, caught on a shelf just visible one hundred and fifty feet [46 metres] below … For three hours we called unceasingly but no answering sound came back. The dog had ceased to moan and lay without a movement. A chill draught was blowing out of the abyss. We felt that there was little hope.[45]

Shocked and grieving, Mawson and Mertz read the burial service for Ninnis and, with only the weakest dogs and little more than a week's rations, began their 'long and painful return journey' back to Main Base, 480 kilometres away.[46] Over the next two weeks, as their rations dwindled, they resorted to killing and eating the weakened dogs. On 8 January 1913, about 160 kilometres from Main Base, Mertz fell ill from exposure and possibly vitamin A poisoning as a result of eating the dogs' livers.[47] Mawson nursed him through his last agonising days. After Mertz's death, Mawson hauled the body on the sledge until, finally, he buried him in the snow and continued on alone across crevasse fields, the shortest route back to Main Base. Mawson speculated that, even if he did not make it back, he could at least find a prominent landmark where a search party might find him. Weakened by exhaustion and starvation, he considered the prospect of making the long trek alone. He wrote:

> For hours I lay in the bag, rolling over in my mind all that lay behind and the chance of the future. I seemed to stand alone on the wide shores of the world—and what a short step to enter the unknown future! My physical condition was such that I felt I might collapse in a moment. The gnawing in the stomach had developed there a permanent weakness, so that it was not possible to hold myself up in certain positions. Several of my toes commenced to blacken and fester near the tips and the nails worked loose. Outside, the bowl of chaos was brimming with drift-snow and I wondered how I would manage to break and pitch camp single-handed. There appeared to be little hope of reaching the Hut. It was easy to sleep on in the bag, and the weather was cruel outside.[48]

His feet continued to worsen as the skin separated from the soles. Each step was agony, and he wrote of how his whole body seemed to be rotting from lack of nourishment. On 17 January, two days after the deadline for the summer sledging parties to return, Mawson fell into a crevasse. Suspended at the end of a rope attached to his sledge, he recalled:

> My strength was fast ebbing; in a few minutes it would be too late. It was the occasion for a supreme attempt. New power seemed to come as I addressed myself to one last tremendous effort. The struggle occupied some time, but by a miracle I rose slowly to the surface.[49]

At the end of January, Mawson discovered a snow cairn marking a food depot established by a rescue party just hours earlier. Replenished with food and the news that the *Aurora* was waiting for him, he managed to reach the ice cave that the AAE men had dubbed Aladdin's Cave. By then, he was just 37 kilometres from Main Base, but another blizzard set in and he was forced to spend a further five days in the hole. Meanwhile, the main group waited anxiously at the Main Base. On 22 January, Frank Stillwell wrote in his diary: 'Still no Mawson'. Five days later, he wrote: 'Still no Mawson. The most optimistic among us now are beginning to have fears not easily calmed'.[50]

↓ Xavier Mertz emerging from Aladdin's Cave, a cavern dug into the ice near Commonwealth Bay that served as a food depot and refuge for sledging parties.

Sledging songs

Out on the ice, the men of the AAE composed 'sledging songs' to help pass the time during long days manhauling heavily laden sledges. The songs also functioned like the traditional sea shanties sung by sailors as they laboured, with sledging parties singing out loud as they marched in time to the rhyming verse. The songs served as a distraction from the monotony and helped to boost morale. They also gave the men an opportunity to celebrate their achievements, and some recorded them in personal diaries or the expedition's newspaper, *The Adelie Blizzard* (see page 77).[51]

Frank Hurley composed the 'Southern Sledging Song' during his traverse to the south magnetic pole. It was sung to the tune of 'Sailing', a popular song composed by Godfrey Marks in 1888, with the chorus:

> *Hauling, toiling, tireless on we tramp*
> *O'er vast plateau, sastrugi high, o'er deep crevasse and ramp.*
> *Hauling, toiling thro' drift and blizzard gale,*
> *It has to be done, so we make of it fun,*
> *We men of the Southern Trail!*[52]

→ Sledging songs helped to boost morale during the long hours hauling heavy-laden sledges across the ice sheet (from left: John Close, Frank Stillwell and Charles Laseron).

Hurley later wrote that the song reflected the difficulties of the journey and the way in which the sledging party had bonded as they endured its hardships:

> As we drew closer [to the main base at Commonwealth Bay] we three, knit together by a great comradeship and affection, our hearts swelling with thankfulness and joy over our deliverance, gave raucous voice to the sledging song that had urged us through many trials and tribulations.[53]

The profound silences and howling blizzards of the Antarctic ice sheet have continued to inspire musical compositions, although such music is increasingly concerned with finding ways to convey the impact of humans in this fragile and distinctive environment. Stuart Greenbaum's choral piece *Antarctica*, for example, was written in 2002 as a lament about rising sea levels in a warming Antarctica, while in 2016 *Antarctica: The Musical* celebrated the work of scientists in a changing polar environment. As musicologist Carolyn Philpott has written:

> Most people will never visit Antarctica. It is an inhospitable place at the margins of our world. But music enables audiences to come to know the continent as a place of both the imaginary and of urgent, practical scientific work.[54]

The second winter

In early February, the hut finally came into view but, to Mawson's dismay, the *Aurora* had already departed, leaving six men to search for his missing party. Sidney Jeffryes, the new wireless operator recruited by Captain Davis, managed to recall the *Aurora* but, despite waiting for the wind to abate, the ship could not return. Finally, Davis could delay no longer and set sail for the Shackleton Ice Shelf to pick up Wild's Western Party. Forced to endure another winter, the seven men remained at Cape Denison until the *Aurora* returned in December 1913. They erected a cross on Azimuth Hill near the Main Base in memory of Ninnis and Mertz, and a Transit Hut to house a theodolite for taking sightings of stars and determining Cape Denison's longitude.

Contemplating the prospect of another long year at Cape Denison, Mawson wrote:

> *Now there were only seven of us; we knew what was ahead; the weather had already given ample proof of the early approach of winter; the field of work which once stretched to the west, east and south had no longer the mystery of the 'unknown'; the Ship had gone and there was scant hope of relief in March.*[55]

To fill the long months ahead, there was still useful scientific work to be done: collecting biological and geological specimens, conducting meteorological and magnetic observations, and writing up the results of the previous year's scientific work. The men also turned their hand to writing and producing a newspaper for their own consumption. Expedition newspapers had their origins in the Arctic during the early nineteenth century, and the idea was enthusiastically embraced by those venturing to the Antarctic. With a print run of one, the articles were spiced with ample in-jokes and obscure references. Their main function was not to disseminate information, but rather to create a shared activity that improved morale and offered relief from the tedium. *The Adelie Blizzard*, produced by Dr McLean, proved a great success, serving to pass the time and boost spirits as they awaited the return of the *Aurora*. Using a Smith Premier typewriter, they wrote an eclectic range of articles from 'news items' and mock advertisements, to competitions, letters to the editor, Births, Marriages and Deaths (featuring the expedition's huskies), book reviews, opinion pieces, sports reports, public

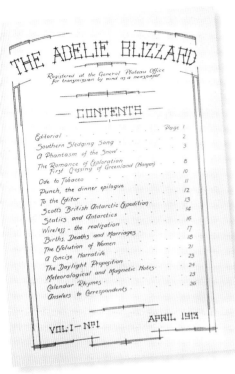

← Creating content for *The Adelie Blizzard*, the first 'real' Antarctic newspaper, served to keep up the spirits of the seven men who remained at Cape Denison for a second year.

notices, scientific reports, amusing accounts of hut life, social issues, poetry, plays and short stories. In all they produced five issues, with a banner declaring that *The Adelie Blizzard* was 'Registered at the General Plateau Office for transmission by wind'. Mawson also had an eye to future publication back in Australia when he contributed a series of educational articles on the 'Commercial Resources of Antarctica'. He claimed it as the first 'real' Antarctic newspaper. Indeed, it was the first to include information transmitted to Antarctica via the new radio wireless technology.

On 15 February 1913, the radio operator Jeffryes, who had arrived in January with the *Aurora*, detected the first coded weather report sent from Macquarie Island to Hobart. The signals became increasingly clear until, a few days later, Jeffryes succeeded in making radio contact with the island. Mawson immediately radioed a message to the Governor-General, Lord Denman, advising him of the situation and the loss of two members of the expedition, along with messages to the families of the two men who had perished. In turn, they received a message of condolence from Robert Falcon Scott's widow, news of Captain Scott's death and that of his four companions during their sledging

→ In February 1913, Mawson sent this radio telegram to Australia's Governor-General with news of the AAE's sledging achievements and the loss of Ninnis and Mertz.

RADIO TELEGRAM

COMMONWEALTH OF AUSTRALIA.
DEPARTMENT OF POSTMASTER-GENERAL

National Archives of Australia

NAA: MP341/1, 1915/2307

↑ Australian audiences flocked to exhibitions and lectures about Antarctic exploration in the early 20th century. Here, Mawson posed with photographs and equipment, including tents and sledges using during his expeditions.

→ A map showing the voyages of the *Aurora* between 1911 and 1914 during the AAE.

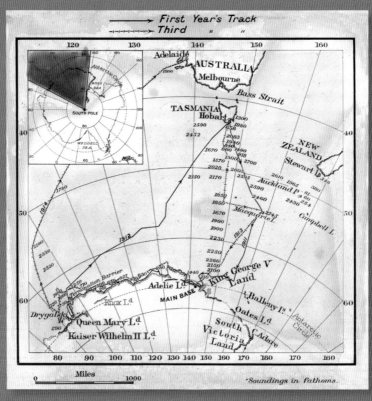

journey to the South Pole, and of the safe return to Hobart of Frank Wild's Western Party aboard the *Aurora*. Jeffryes' skill with the radio played a crucial role in maintaining morale during that second year. Nevertheless, Jeffryes, the newcomer responsible for maintaining communications with the outside world, became increasingly isolated from his companions. As his mental health rapidly deteriorated, he became paranoid and violent.[56] Mawson too, having returned 'emaciated and broken' from his sledging journey after losing Mertz and Ninnis, became the focus of resentment among some of the men.[57] As winter progressed, the men also had to contend with unusually heavy snowfalls and the incessant blizzards. Mawson, Madigan and Hodgeman made a short sledging journey to Mount Murchison to recover instruments left there by the summer sledging parties. Finally, on 12 December, the *Aurora* came into view. 'There were wild cheers down at the Hut when they heard the news,' Mawson recalled. After two long years of isolation, the men boarded the ship: 'At dinner we sat down reunited in the freshly painted ward-room, striving to collect our bewildered thoughts at the sight of a white tablecloth, Australian mutton, fresh vegetables, fruit and cigars.'[58]

Within a year of his return, Douglas Mawson had received a knighthood and was beginning a series of lectures, illustrated with Frank Hurley's photographs. Payment from the lectures would help to pay off the debt he still owed on the expedition.[59] Australian newspapers reproduced a selection of Hurley's photographs under the banner 'Mawson's expedition to the ice-bound south' but, when the First World War dramatically intervened, public attention soon shifted away from Antarctic exploration.[60] Nevertheless, the AAE had achieved a great deal. Throughout the two arduous years, the expeditioners had charted much of the eastern Antarctic coastline and made detailed observations in magnetism, geology, biology and meteorology sufficient to fill 22 volumes of scientific reports. Meanwhile, Mawson published his personal account of the expedition in 1915 while convalescing from his ordeal and, in subsequent decades, this first Australian-led expedition would receive international acclaim. The final volumes of the AAE's *Scientific Reports* were not published until 1947, 11 years before Mawson's death. According to the British geographer and meteorologist J. Gordon Hayes, Mawson's expedition was 'the greatest and most consummate expedition that ever sailed for Antarctica'. He wrote:

MAWSON'S EXPEDITION TO THE

The fo'castle head of the Aurora covered with frozen sea spray after a blizzard.

This cross is erected on a high rocky cliff close to the winter quarters to the memory of Lieut. B. E. S. Ninnis and Dr. X. Mertz, who lost their lives while sledging. On the tablet are the words:—"Erected to commemorate the supreme sacrifice made by Lieut. B. E. S. Ninnis, R.F., and Dr. X. Mertz in the cause of science.—A.A. 1913."

Evening at the Anchorage, Commonwealth Bay, Adelie Land, December, 1913, showing a huge overhanging

The ice barrier, showing the impregnable cliffs where the inland ice meets the frozen sea. This photograph was taken immediately

Crevassed ice near the winter quarters, Adelie Land. Frequently miles of this country has to be crossed whilst sledging. These photos. were taken by Mr. Frank Hurley, who accompanied the expedition.

THE RETURNING MEMBERS OF THE LAND PARTIES. Back Row.—L. R. Blake, A. L. McLean. Middle Row.— F. Ainsworth, C. T. Madigan, E. Bage, Dr. Mawson, C. A. Sandell, P. E. Correll. Front Row.—A. J. Hodgeman,

BOUND SOUTH.

Penguin and sea elephant life at Macquarie Island. The adult bull elephants frequently weigh from 2½ to 3 tons.

Dr. Mertz emerging from Aladdin's Cave, a cavern excavated in the glacier ice, and used as a food depot. Observe the canvas flap, used as a door, thrown back and the two food bags, each containing a fortnight's supply for three men.

Antarctic petrels nesting at Cape Hunter, Adelie Land. Little has hitherto been known of the nesting habits of these birds, and the observations made by the biologists of the Australian Antarctic Expedition will be of great value.

The expeditions of Scott and Shackleton were great, and Amundsen's venture was the finest Polar reconnaissance ever made; but each of these must yield the premier position, when fairly compared with Mawson's magnificently conceived and executed scheme of exploration.[61]

← In March 1914, Australian newspapers reproduced a selection of Hurley's images from the AAE to showcase the beauty and perils of the otherworldly region to the south.

British, Australian, New Zealand Antarctic Research Expeditions

It was 13 January 1930 when Mawson set foot once more on the Antarctic ice. It had been 17 years since the rescue of the remaining members of the AAE from Cape Denison, and the world was in the midst of the Great Depression. Since the end of the First World War, Mawson had been urging the Australian government to support another Antarctic expedition in order to complete the geographical work he had begun. An Imperial Conference held in London in 1926 concluded that further exploration and scientific research was needed if Britain was to consolidate its territorial interests in Antarctica. In 1927, the Australian government invited Mawson to organise and lead the British, Australian, New Zealand Antarctic Research Expeditions—BANZARE.[62]

Over two summer expeditions in 1929–1930 and 1930–1931, BANZARE expeditioners undertook an extensive scientific program in the Southern Ocean, which, Mawson noted, would enable Australia to develop its fisheries in Antarctic waters. His scientific staff included biologists, zoologists, a meteorologist, an ornithologist, two aviators and a hydrologist. In addition to ocean research, the expeditioners conducted flights to chart previously unknown areas of the continent in preparation for a territorial claim for the British Empire covering more than one-third of Antarctica.

With ice closing in on the *Discovery*, Captain Davis made the decision to land on a barren island of rock, coming as close to the coast as he dared. A group of ten expeditioners rowed to shore by boat and immediately climbed to the summit of a low hill. At precisely midday, the BANZARE expeditioners stood as the Union Jack was unfurled above a cairn hastily made of loose rocks, and Mawson declared British sovereignty over a portion of Antarctic ice. As it happened, the canister containing the proclamation had already been buried in the cairn so Mawson, with promptings from Hurley and Moyes, recited the words from memory.

> *In the name of His Majesty King George the Fifth of Great Britain, Ireland and the British Dominions beyond the Seas, emperor of India. Whereas I have in command from His Majesty King George the Fifth to assert the sovereign rights of His Majesty over British land discoveries met with in Antarctica. Now, therefore, I Sir Douglas*

Mawson do hereby proclaim and declare of all men that, from and after the date of these presents, the full sovereignty of the territory of Enderby Land, Kemp Land, Mac Robertson Land, together with off-lying islands as located in our charts constituting a sector of the Antarctic Regions lying between Longitudes 73° East of Greenwich and 47° East of Greenwich and South of Latitude 65°, vests in His Majesty King George the Fifth, His Heirs and successors for ever. Given under my hand on board the Exploring vessel 'Discovery' now lying off the coast of this annexed land, in Latitude 65°50' S. Longitude 53°30' E. The Thirteenth Day of January, 1930.[63]

As the men removed their hats and sang 'God Save the King', Frank Hurley filmed the occasion for posterity. So began Mawson's mission to take possession of 5.9 million square kilometres, about 42 per cent, of Antarctica as British sovereign territory. Within three years, Britain would transfer its Antarctic territorial interests to Australia under the *Australian Antarctic Territory Acceptance Act 1933*, which came into effect in 1936.[64]

↓ Raising the Union Jack on Proclamation Island during the first BANZARE, 13 January 1930.

The island lies buried in mist and fog, amidst the turmoil of the great rolling seas which, in those latitudes, sweep unchecked around the Globe.

SIR DOUGLAS MAWSON[1]

Island

Stories from the subantarctic

← Looking north
across the isthmus on
Macquarie Island.

Dr Jaimie Cleeland sits quietly in the tussock grass at the southern tip of Macquarie Island. She is watching two male wandering albatrosses (*Diomedea exulans*) sitting side by side on their nests, each nursing an enormous egg. They fledged from this spot eight or so years earlier, embarking on a circumpolar journey covering many thousands of kilometres in these 'albatross latitudes' before returning to their birthplace to breed.[2] Their female partners, each having laid a 500 gram egg, have headed out to sea where they will spend several weeks foraging to replenish their energy while the males care for the precious eggs.[3] From time to time, one of the males looks down at his egg and sings to it in a deep, guttural voice.[4]

'Macca'

Cleeland came to Macquarie Island, commonly known as 'Macca', in 2011 for the first of three consecutive summers to study the island's four species of albatross. She spent most of her time in the field, regularly walking the length of the 34 kilometre–long island from the 'albatross' field hut at Hurd Point in the south to the station in the north, in order to inspect the albatross colonies. 'Walking the length of the island in a day is no easy feat,' she recalls.

> Initially, you have a huge climb to get up off the coast onto the plateau. Plateau trekking is pretty straightforward, until you reach sections of the track like Windy Ridge, where the wind funnels up from the west coast and can often reach well over 50 knots [93 kilometres per hour]. Dropping down into Green Gorge you get a beautiful view across the basin with the sun beaming off the escarpment; however, there are no dry feet as you squelch your way across the tarn below. The last challenge when heading north is Doctors Track, leading down into station. For tired knees, going downhill can be a real struggle, especially on this muddy, steep track.[5]

→ Jaimie Cleeland negotiating Macquarie Island's steep terrain during her fieldwork, with southern rockhopper penguins (*Eudyptes chrysocome*).

The nests of the light-mantled albatrosses (*Phoebetria palpebrata*) are perched on the steep sides of the island's rugged sea cliffs, high above the surrounding ocean, while wandering albatrosses generally nest in the grassy areas above the cliff. In order to take flight, these majestic birds will waddle to a higher point and extend their giant wings, allowing a powerful updraft to lift and propel them into the turbulent winds of the Southern Ocean. For albatrosses, the process of nest-building, mating and hatching eggs must be undertaken in the brief summer months.

Cleeland's own routine is frenetic: rise early, get to the first study site of the day as soon as possible, crawl slowly to each nest and lift the feathers on the side of the nesting adult just enough to spot the identifying band, record the number of eggs and adults in each colony and move on to the next site. She has to find and record all the eggs before the chicks hatch, but trying to read a band on the leg of a nesting male is an exercise in patience. The male wandering albatross has the largest wingspan of any flying bird in the world. It can measure up to 3.5 metres across, so attempting to lift its feathers to get a

look at its egg can 'get you a good whack'. The scientist moves slowly, keeping her eyes averted. She will return to each colony several times each season to record the leg bands of the females, again when the eggs are hatched and, finally, in the weeks between October and December when the chicks fledge and are preparing for their own circumpolar journeys. If her timing is right, she will find the fledglings standing on their nests, beating their wings to build up muscle strength before shuffling a metre or so away, testing their balance as the wind fills their enormous wings and lifts them off the ground. Soon they will be ready to stand at the edge of the cliff and take flight, with their heavy bodies borne by the powerful forces of the Furious Fifties.

← An albatross chick on Macquarie Island.

→ An adult albatross nesting on Macquarie Island.

The most wretched place

During her walks, Cleeland sometimes comes across the haunting remains of the island's early oil-harvesting industries. Macquarie Island was first sighted in 1810 by the British-born Australian sea captain Frederick Hasselborough aboard the *Perseverance* as he searched for fur sealing grounds. He named it after Lachlan Macquarie, then governor of New South Wales. Like its subantarctic island cousins, Macquarie's isolated, ice-free beaches are important summer refuges and breeding grounds for huge numbers of Southern Ocean birds and marine mammals. With news of Hasselborough's discovery, sealing gangs converged on the island and, within a decade, they had slaughtered an estimated 120,000 fur seals to meet the world's voracious demand for animal fur. With the near extinction of fur seals, the sealers turned to harvesting the oils of the island's southern elephant seals and king penguins (*Aptenodytes patagonicus*).[6] It was a gruesome business,

↑ Rusting remnants of the sealing era can still be found on Macquarie Island.

and Macquarie Island gained a reputation as a forbidding place. Australia's colonial administrators deemed it too bleak and desolate even for a penal settlement. When Captain Douglass, master of the vessel *Mariner,* visited the island in 1822 to load barrels of elephant seal oil bound for London, a Sydney newspaper reported his observations:

> As to the Island … it is the most wretched place of voluntary and slavish exilium that can possibly be conceived: nothing could warrant any civilized creature living on such a spot, were it not the certainty of industry being handsomely rewarded; thus far, therefore, the poor sealer, who bids farewell, probably for years, to the comforts of civilized life, enjoys the expectation of ensuring an adequate recompence for all his dreary toils. As to the men employed in the gangs … they appear to be the very refuse of the human species, so abandoned and lost to every sense of moral duty.[7]

With its location at latitude 54° S, about halfway between New Zealand and Antarctica, the remote island in the midst of the Southern Ocean received few human visitors at first, apart from sealers and shipwreck survivors. By the end of the nineteenth century, it began to assume strategic importance as a convenient resupply point for vessels embarking on, and returning from, hazardous voyages to Antarctica.[8] The Australasian Antarctic Expedition (AAE) visited Macquarie en route to the Antarctic continent in 1911. Part of the AAE's mission was to establish an Australian station on Macquarie Island, and to leave a small party there for a year, under the leadership of George Ainsworth from the newly formed Bureau of Meteorology. As Captain Davis guided the *Aurora* into Caroline Bay, he wrote:

> Thick tussock-grass matted the steep hillsides, and the rocky shores, between the tide-marks as well as in the depths below, sprouted with a profuse growth of brown kelp. Leaping out of the water in scores around us were penguins of several varieties, in their actions reminding us of nothing so much as shoals of fish chased by sharks. Penguins were in thousands on the uprising cliffs and from rookeries near and far came an incessant din. At intervals along the shore sea elephants disported their ungainly masses in the sunlight.[9]

Over the next year, the party would map the island, study its distinctive flora, fauna and geology, and make regular meteorological observations.[10] Their first task was to erect wireless masts on the top of a hill at the northern end of the island that were used to relay messages between Mawson's main expedition base at Commonwealth Bay and the Australian mainland. Mawson was enthralled by the island and its prodigious wildlife.

> *In outline the Island is long and narrow, being over twenty miles [32 kilometres] in length, and with a fairly uniform width of three to three and a half miles [5 to 6 kilometres]. The rocks are chiefly volcanic, and rise rather abruptly as a ridge from the floor of the deep sea. The salient physiographic feature is the steepness of the land; the summit forming a miniature plateau having an elevation above sea-level of 1,000 feet [305 metres] or thereabout.[11]*

↓ **The AAE Macquarie Island party in 1911–1912: (from left) Charles Sandell, wireless operator and mechanic; George Ainsworth, leader and meteorologist; Arthur Sawyer, wireless operator; Harold Hamilton, biologist; Leslie Blake, geologist and cartographer.**

↑ The wireless engine hut established on Wireless Hill, Macquarie Island, during the AAE.

→ The wireless operating hut was central to the AAE's mission on Macquarie Island, enabling the first wireless messages to be relayed between Mawson's Main Base at Cape Denison and mainland Australia.

While the others laboured to set up the wireless station, the expedition's photographer, Frank Hurley, and biologist Charles Harrisson, accompanied by one of the few sealers still on the island, trekked back to Caroline Cove to retrieve a camera lens that Hurley had lost during the first landing on the southern end of the island. After four days, the group trudged into the station, muddy and exhausted, hauling a large wandering albatross they had killed along the way to add to the expedition's natural history collection.[12] When Mawson and his Main Base party sailed for Antarctica, the five remaining men set about building their own small living hut at the base of Wireless Hill, where the winds blew almost continuously. They would remain there until the *Aurora* returned the following summer.

→ Arthur Sawyer and Charles Sandell, who operated the wireless at night when reception was best, would often bunk down in the operating or engine hut on Wireless Hill rather than risk returning in darkness to the expedition hut on the isthmus.

The subantarctic territories: Macquarie Island

Macquarie Island (Macca) is a rare place on Earth. It is the only island created entirely from oceanic crust and rock squeezed up from deep within Earth's mantle, and it offers a unique geological window into the planet's ancient past. The island probably formed between 30 and 11 million years ago, six kilometres below Earth's oceanic crust. As the seafloor spreading stopped, the crust began to compress, squeezing rocks upward to form a ridge. About 600,000 years ago, the ridge emerged as an island above the surface of the Southern Ocean.[13] Macquarie is also an ecological wonderland where, as on its subantarctic island neighbours, huge numbers of Southern Ocean birds and mammals seek refuge: feeding, breeding and nursing their young on its isolated shores.

Between 1820 and 1930, nine expeditions are known to have visited Macquarie Island. These include von Bellingshausen's Russian expedition in 1820, Charles Wilkes's United States Exploring Expedition in 1840, Borchgrevink's *Southern Cross* expedition in 1898, Scott's *Discovery* expedition in 1901 and Shackleton's *Nimrod* expedition in 1909. In 1959, two Australians—marine biologist Dr Isobel Bennett and intertidal ecologist Hope McPherson—were invited to share a cabin with British

← King penguins at Lusitania Bay, Macquarie Island, 1984.

biologist Mary Gilham and biological secretary Susan Ingham on the Danish polar vessel *Thala Dan* during its annual relief and resupply voyage for the personnel based at Macquarie Island. As the chartered naval ship approached the island, Dr Bennett wrote, 'First impressions are the ones which always remain … high forbidding cliffs looming eerily out of the misty dusk, with an echoing shout of human voices borne on the wind' as station staff welcomed the arrival of the relief vessel after 'long, dreary months'. Bennett and her colleagues were the first women to be allowed to join an Australian National Antarctic Research Expedition (ANARE). On stepping ashore, she saw the signs of the 'wanton and unrestricted destruction of native species', a legacy of rats and other feral animals introduced to the island by earlier inhabitants. She wrote:

> *The Island's seashores are far away from the normal haunts of man. It may be that while Macquarie remains a sanctuary, the animals living there will be left to live out their lives as they had done for thousands of years before man's arrival. But with the enormous problems facing an ever-growing world population, with its increasing demands for more and more food, future generations may again take toll from this far-off land.*[14]

In 1997, 50 years after the first ANARE and more than 60 years after being declared a wildlife sanctuary, Macca was inscribed on the World Heritage List in recognition of its outstanding natural features. Two years later, the Australian government created the Macquarie Island Marine Park to protect more than 16 million hectares of the surrounding ocean as a habitat for threatened species of penguins, fur seals, southern elephant seals and albatrosses. Nevertheless, the 'wanton and unrestricted destruction' that Dr Bennett described in 1959 still plagued the island. Finally, some two centuries after sealers introduced rabbits, rodents and cats to the island, a massive feral animal eradication program began, using hunters and specially trained dogs. It took nine years to clear the island of the unwanted creatures.[15] When the Australian government announced that the station on Macquarie Island would be closed in 2016, there was a public outcry. The plan changed, and a decision was made to retain and renovate the station. Macca, despite its bleak weather and rugged terrain, continues to be a mecca for scientists and tourists alike.

This little island

At the time of Mawson's visit to Macquarie Island, a thriving penguin oil business was being run out of Invercargill on New Zealand's southern coast by Joseph Hatch, a British chemist who had migrated to New Zealand in 1862 and served as a member of the New Zealand Parliament in the 1880s. Hatch had begun his ventures in animal processing with bone milling, rabbit-skin exporting and soap and glycerine manufacturing, before turning his hand to fur sealing on New Zealand's subantarctic islands. In 1873, the New Zealand government introduced closed periods for sealing in an effort to halt the rapid decline in the animals' numbers. After seal skins were discovered on board one of his vessels outside the sealing season, Hatch relocated his operations to Macquarie Island, which was by then administered by Tasmania and had no such restrictions. There, he developed his business processing elephant seals and experimenting with extracting the oil of king penguins using a digester works.[16]

By the 1890s, newspapers were reporting that Hatch's sealers were herding live penguins into the digesters, drawing public outcry in Australia and Britain. Hatch turned to public lectures to promote his venture. During one of these, at the Princess Theatre in Dunedin, he made the case for his enterprise with a presentation titled 'The Macquaries: beyond the reach of civilisation'. After presenting images displaying the island's beauty, he showed the millions of king penguins that gathered on the island to breed, ridiculing the idea that his operations were causing them to become extinct, since many more chicks fell victim to sea hawks and giant petrels than to his harvesting activities.[17]

Hatch continued to promote the natural beauty and resources of Macquarie Island in public lectures after moving to Hobart in 1912. By the time Mawson's men arrived, Hatch was sending a party to the island each year to kill penguins and elephant seals for their oils.[18] Leslie Blake, the expedition geologist and cartographer, described how a couple of men would drive 2,000 birds at a time into a netted enclosure where they would select the fat one-year-old chicks, knock them on their heads and pack them into huge steam boilers to extract the prized oil.[19] When Mawson observed the oil-harvesting operations, Hatch was leasing the entire island from the Tasmanian government, but the public perception that penguin oiling was a cruel business never dissipated, and Hatch

↑ Royal penguins and elephant seals on Macquarie Island,
photographed by Frank Hurley during the AAE's visit in 1911.

was still defending himself at public meetings in 1922 while campaigning for election to the Tasmanian parliament at the age of 85. Hatch's lease on the island expired in 1920.[20]

Mawson was concerned that unregulated harvesting would spell disaster for Macquarie Island's wildlife, as it had for its fur seal populations, and began campaigning in earnest for the island to be declared a bird and animal sanctuary in recognition of its distinctive qualities. As he argued during a lecture in 1919:

> *This little Island is one of the wonder spots of the world. It is the great focus of the seal and bird life in the Australasian sub-Antarctic Regions, and is consequently of far greater significance and importance in the economy of that great area than its small dimensions suggest. This being so, it behoves those responsible for its good keeping to see to it that the animals resorting thereto are properly protected against any possibility of extermination … If we totally disregard the value to science of perpetuating this life for study and observation in the future; if we take no count of the pleasure which such an island of animal life must afford future generations of the human race who, alas, are destined to live in a world more and more robbed of its former varied fauna and flora, if we are not perturbed by the prospect of an ocean devoid of the bird life so welcome to the ocean voyager; if we give no thought to the economic possibilities of exploitation in future times; and finally if the wholesale destruction of animal life leaves us unmoved, then we may proceed to continue with the unchecked slaying for the pence that it returns and with no thought for the morrow.[21]*

Mawson's campaign eventually succeeded and, in 1933, the Tasmanian government revoked all sealing licences and declared Macquarie Island a wildlife sanctuary.[22] In the following decade, the island became home to the newly formed ANARE with a station established at the foot of Wireless Hill on the windswept isthmus at the island's northern end, near the site of the AAE's 1912 hut. The first ANARE party, comprising 13 men, arrived on 21 March 1948 to begin the scientific research and monitoring programs on the island that continue to the present day.

→ Elephant seals and king penguins photographed on Macquarie Island by Peter Dombrovskis in 1984.

The subantarctic territories:
Heard and McDonald islands

The black specks of rock known as Heard and McDonald islands emerge from a huge plateau that is otherwise underwater, breaking the surface of the vast Southern Ocean 4,100 kilometres southwest of Cape Leeuwin on the Australian mainland. They form one of 20 or so tiny island groups that lie in the stormy circumpolar ocean between latitudes 40° S and 60° S, and are among the most isolated fragments of land in the world.[23] When the Australian polar explorer and writer John Béchervaise landed on Heard Island with the 1953 ANARE, he was as excited at the prospect of skiing and dog sledging (the party had brought a team of huskies with them) as his scientific work. He wrote:

> I shall not forget my first-footing on Heard Island … We walked up amongst
> the great mossy tussocks and over the extraordinary lava flows, which in places
> look as though they might have cooled last week, to the cliffs and overlooking
> the cove. I found the landscape fascinating. A few shades of green cushion
> plant … hardy enough to survive the deep winter snow growing in windswept
> arcs, the astonishing rope-like twisted lava, numerous sea elephant cows, quite
> torpid, a tribe or two of crested penguins on the cliff-top, and some utterly
> fearless cormorants.[24]

In 1955, after eight years of operation, the Australian government decided to close the research station on Heard Island to focus its efforts on establishing one on the Antarctic continent. In the summer months each year, ANARE scientists continued to visit the island to undertake seismic, meteorological, glacial and wildlife research. Between 1947 and 1971, successive expeditions also circumnavigated it, climbed the volcano—Big Ben— and made the first recorded landing on McDonald Island on 27 January 1971. Unlike the more accessible subantarctic islands, Heard Island has largely avoided the ravages of predators such as cats and rats, although its prolific colonies of elephant seals did not escape the sealing gangs, whose long-abandoned stone huts and rusted iron rendering pots can still be found. In 1997, the Territory of Heard and McDonald Islands, once described by Sir Charles Wyville Thomson, chief scientist of the *Challenger* expedition, as 'the most desolate spot on God's earth', was inscribed on the World Heritage List.[25] These isolated islands, with their inhospitable climate, are rarely visited even today, and are considered to be among the most biologically pristine areas in the world.

↑ A light amphibious resupply cargo vessel transporting gear to Heard Island in 2014.

Fourteen men

Some 5,000 kilometres to the west of Macquarie Island, another group of Australians were at work on Australia's other subantarctic territory. Arthur Scholes was awestruck by his first glimpse of Heard Island, emerging from the ocean surface far to the west of Australia's southern coastline. From the deck of the Royal Australian Navy's Landing Ship Tank (LST) 3501, he described his first impressions:

> *The scene ahead was unforgettable. In my mind I had been building up a picture of a grim, desolate country, obscured by fogs and mist. The first glimpse of the island was possibly the best view we had. In the dead calm blue of the sea, there lay the island, alone and majestic, like a great white iceberg.*[26]

↓ ANARE members inspecting the inhospitable coast of Heard Island in 1948. It would be their home for the next year.

Heard Island is defined by an active volcano known as Big Ben. At 2,745 metres, it is the highest mountain in Australian territory, and one of Australia's only two active volcanoes. Beyond the island's shingle and sand beaches, the slopes of Big Ben are permanently covered with glaciers. Nearby McDonald Island is even smaller, and it too has its own active volcano, which last erupted in 1992, changing the shape of the whole island and doubling its size. The nearest neighbour to these remote outposts is the French subantarctic territory of Kerguelen Islands, 450 kilometres to the northwest. Unlike the Kerguelen Islands, however, Heard and McDonald are relatively unknown and rarely visited. In 2005, the Australian Antarctic Division estimated that Heard Island had received just 240 shore-based visits since 1855; McDonald Island had just two.[27]

An American sailor, John Heard, claimed to be the first to sight Heard Island in 1852. He bestowed his name upon it and, a year later, William McDonald followed suit when he sighted a scattering of small islands nearby. Within 30 years, sealers had found their way to the nutrient-rich waters around these isolated shards in the subantarctic waters of the Southern Ocean. The British *Challenger* and German *Arcona* expeditions approached Heard Island in 1874, but neither vessel lingered in the poor weather conditions characteristic of the island. As national rivalry to reach the South Pole intensified, Britain laid claim to Heard Island and the nearby McDonald group in 1910. Expeditions from other nations also visited the island to undertake scientific research, including the German Antarctic Expedition in 1902, a French geological expedition in 1928 and Douglas Mawson's British, Australian, New Zealand Antarctic Expeditions (BANZARE) in 1929, although the isolation and notoriously bad weather deterred close inspection of the tiny islands. During the Second World War, German ships took refuge there and Britain, keen to protect its strategic interests in the region, urged Australia and South Africa to assert sovereignty over the islands closest to them. In early 1947, Britain quietly transferred its sovereignty over Heard and McDonald islands to Australia.[28]

Scholes was a member of the newly created ANARE, a program aiming to establish permanent Australian Antarctic stations and support scientific and exploratory work in the region. This first mission in the subantarctic would ensure that the Australian flag was flying on these tiny, windswept dots in the Southern Ocean.[29] The LST 3501 headed for Heard Island in November 1947, while a second vessel, the *Wyatt Earp*, continued on to Commonwealth Bay. Here, the landing party would occupy Mawson's old Main Base Hut and begin a reconnaissance of the coast of George V Land to find a suitable site for Australia's first permanent station on the continent. The two vessels were then to meet up at Macquarie Island. By establishing a physical presence, Australia was asserting its permanent presence in the region and, in the tradition of Mawson's AAE, the mission would be filmed for posterity.[30]

On this tiny volcanic island about halfway between Africa and Antarctica, in one of the most isolated, windswept and unforgiving environments on Earth, Scholes would spend a year with 13 others and limited resources, mapping Heard Island's features, establishing a meteorological station and recording observations of tides, geology and glacial history. The landing party arrived with only a rough map compiled by sealers in the nineteenth century and notes from four earlier voyages that had briefly visited Atlas Cove on the island's northern coast. A week into their stay, their amphibious aircraft was wrecked by wild winds as they attempted to take aerial photographs of Big Ben. Indeed, Big Ben—and the island's weather—would dominate the lives of that first ANARE party.

↓ ANARE members setting up a weather station on this tiny volcanic island in the midst of the Furious Fifties, 4,100 kilometres southwest of Australia.

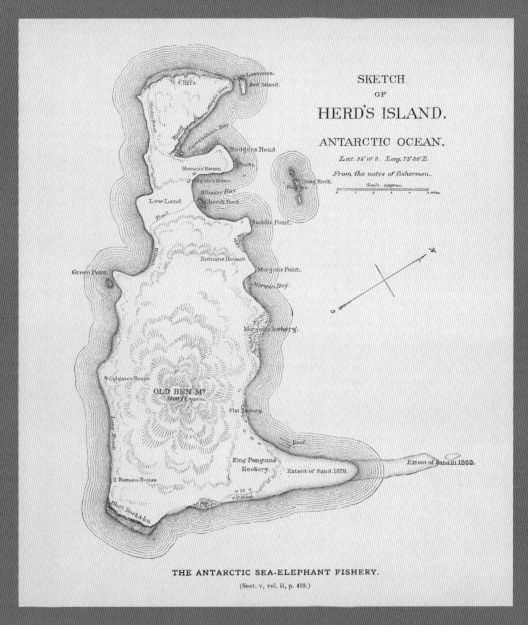

↑ The United States Commissioner of Fisheries, G. Brown Goode, produced this sketch
of Heard Island in 1887, probably based on information provided by early American
sealers and the British *Challenger* expedition, which stopped at the island in 1874.

Windy city

After a failed attempt to land in heavy swells at Spit Bay at Heard Island's southern extremity, Group Captain Stuart Campbell decided to establish the main station in Atlas Cove, close to the old Admiralty Hut built in 1927 by the British government as a refuge for shipwreck survivors and briefly occupied in 1929 by Mawson's BANZARE party. Landing operations were constantly hampered by gale-force winds. When the ship was forced to relocate to deeper water for safety, some of the expeditioners spent their first night on the island. Campbell camped with one group at a possible station site on the low isthmus between Atlas Cove and West Bay, but they abandoned the idea and named the site Windy City. The other group weathered the storm in Admiralty Hut. Scholes recalled:

> The whole party gathered in the [Admiralty] hut for supper.
> They brewed hot cups of kai on the chip heater. Smoke
> and fumes poured from the stove, A chimney, made from
> ration tins and piping, led to the centre of the pyramid roof.
> The high wind outside choked this outlet. The atmosphere
> in the hut soon resembled that of a Chinese joss-house.
> Suffocating smoke forced the men to open the door.
> Cooked tinned sausages and peanuts completed the meal.[31]

The expeditioners and ship's crew worked long hours between storms until every muscle ached, unloading equipment, machinery, fuel and food for the year. There were some lighter moments, despite the gruelling conditions. Scholes recalled one occasion when 'Doc' Gilchrist was unloading his motorcycle from the ship:

→ The arrival of the first ANARE at Heard Island in December 1947 marked the beginning of Australia's postwar ambition to establish permanent research stations.

We stood and cheered encouragement when the 'Doc' appeared at the head of the ramp, astride the throbbing bike. Down he bumped, splashing through the water to the rocks, disdaining assistance. He rode the jumping machine up the slope to the flat, to a thunderous chorus of 'Whoopees! Yoohoos! And Yippees!' He was the first man to ride a motor cycle in the Antarctic.[32]

← **ANARE members of the Heard Island party take a break between unloading shifts in 1948.**

As they continued their labours, Scholes recalled a mysterious change of wind direction that becalmed the battered landing ship. As thick fog enveloped the island, he described the curious effect on the seabirds when the main deck lights were turned on to guide the work-boat in the gloom.

> *Hundreds of small 'whale' birds, light blue, like large butterflies, fluttered round the lights, settling on the deck, boats, hatchways and ventilators. Seamen hauling in the work-boat were careful not to tread on them, there were so many. Most of the birds disappeared when the lights were extinguished. Ivanac found two 'stragglers' outside his cabin. He took them inside. They slept the night in his sea-boots.*[33]

Such moments of calm, however, were fleeting. When the party experienced the full force of a hurricane during their second week of landing operations, even the ever-circling seabirds were forced backwards midair. With gusts reaching more than 120 miles per hour [190 kilometres per hour], those sheltering in Admiralty Hut swore that the tiny building lifted off the ground, straining against its guy wires.[34] Outside, the ship began dragging its anchors and the aircraft was ripped from its moorings, the wreckage scattered to the four winds. By Christmas Day, the hurricane had eased and the expeditioners were treated to a snowy Christmas. They rose early, singing Christmas carols interspersed with 'Waltzing Matilda', while they finished dragging the conglomeration of stores, food, fuel drums and hut materials up from the beach along a track they christened Burma Road. The chief engineer was unimpressed: 'Call this glamour! ... All those people who talk about the glamour of a White Christmas should be shipped to Heard Island to see it for themselves!'[35]

In the days to come, Scholes found solace from the island's air of sullen harshness in Big Ben's magnificent white dome. 'There was little beauty in the gaunt grey rocks,' he wrote, 'the barren flat and grim precipitous coastline. But, despite all that, there was something of almost indefinable loveliness about it.'[36] On Boxing Day 1947, the small party gathered on a bleak ice-free area under the silent gaze of Big Ben to raise the Australian flag, while Campbell read a proclamation declaring this tiny shard of land to be Australian territory. It was a modest celebration, conducted thousands of kilometres from Australia's shores

before an assembly of elephant seals lying among the tussock grasses, but it signalled Australia's intention to have a permanent presence in the high southern latitudes. The first Australian National Antarctic Research Expedition had begun.[37]

↖ Admiralty Hut at Atlas Cove on Heard Island, built in 1927 by the British government as an emergency hut for distressed mariners, was briefly occupied by Mawson's BANZARE expedition in 1929 and ANARE from 1948.

→ Volcanic activity has been observed on Big Ben since the 1880s. Satellites detected an eruption in October 2012, later confirmed in this photograph produced by NASA Earth Observatory showing fresh lava flow on Mawson Peak.

Here was a new world, flawless and unblemished, into which no human being had ever pried. Here were open secrets to be read for the first time. It was not with the cold eye of science alone that we gazed at these rocks—a tiny spur of the great unseen continent; but it was with an indefinable wonder.

SIR DOUGLAS MAWSON[1]

Territory

← Antarctica's glaciers flow from the ice sheet out to the margins of the continent where giant icebergs calve into the surrounding ocean. If the entire Antarctic ice sheet were to melt, global sea levels would rise by around 60 metres.

Dr Phillip Law, leader of the Australian National Antarctic Research Expedition (ANARE), could just make out a slope of rock, a grey smudge among the all-encompassing white ice. It was 1954, and the *Kista Dan* was nudging its way into the sea ice surrounding the continent. Beyond the rocky coastal edge, the Antarctic plateau stretched out to the horizon, broken only by the top of Mount Henderson about 16 kilometres inland and the distant peaks of the Casey, David and Masson ranges in the Framnes Mountains. Law's mission was to locate an ice-free site that would be suitable for building a permanent station. It would be the first built by any nation south of the Antarctic Circle.[2] At the time, the Australian government's priority was to carry forward Mawson's scientific legacy by establishing a base in Antarctica that would serve to consolidate its sovereignty on the ice continent. The Australian government was sensitive to potential challengers in its newly acquired territory, with Norway actively exploring the coastline during the 1930s and the United States beginning to show interest in the 1940s. Australia's new polar 'empire' was still largely unexplored, but the vast territory held the promise of rich pickings, both from minerals buried in its rocky coastal mountains, and from the whales and fish in its surrounding waters. Professor Patrick Quilty called it 'reconnaissance science'.[3]

Reconnaissance science

The prospect of undertaking a scientific program in such a vast and inaccessible territory was daunting. Australia's territorial claim encompassed more than five million square kilometres of the continent and extended to the far-flung subantarctic islands of Macquarie, Heard and McDonald. Some 5,000 kilometres of Antarctica's coastline remained to be mapped, and ships had only a small window in summer during which they might gain access to the coast while the surrounding sea ice was at its minimum extent. The physical challenges of undertaking scientific research in 'Greater Antarctica' were compounded by the fact that Australia, with no permanent ice sheet of its own, lacked scientists with expertise in glaciology.[4] As Law recalled, preparing the logistical support for ANARE's field research programs was extraordinarily complex.[5]

↓ *Kista Dan*, the Danish icebreaker under charter to the Australian government, forges through pack ice at the approach to Sandefjord Bay, Antarctica, in 1955 en route from Heard Island to Mawson research station. A seal hunting party disembarked at Prydz Bay to obtain fresh meat for the 15 huskies aboard the ship.

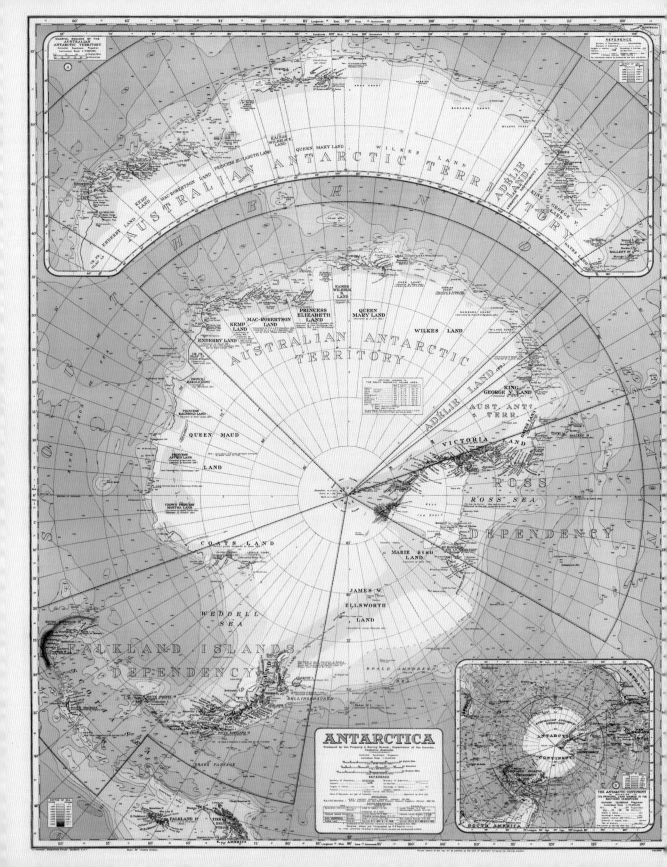

ANTARCTICA

Law's first attempt to reach the continent in 1947 had been foiled by mechanical problems. The aging vessel *Wyatt Earp* was not up to the task and returned to Australia for repairs. The lack of a suitable ship continued to plague the ANARE program, affecting the annual relief voyages to the stations on Heard and Macquarie islands. When the *Kista Dan*, a new Danish icebreaker, became available for charter in 1953, Law jumped at the opportunity and immediately began preparations for a small-scale Antarctic expedition to locate a suitable site for a station in the following year.[6] The station would be named Mawson, in honour of Sir Douglas Mawson. Selecting a suitable site on that vast territory was no easy feat. Law, with the help of Mawson, had pored over aerial photographs taken by a United States reconnaissance aircraft in 1946–1947 as part of Operation Highjump. The US Navy had undertaken this massive air and sea operation to map extensive areas of the Antarctic continent—much of it never before seen—and determine the feasibility of establishing US bases in the Antarctic. They identified several potential sites, but Law finally chose a rocky outcrop on the coast, with a cluster of offshore islands and the Framnes Mountains beyond.[7]

The site seemed ideal. On a coastline dominated by ice, it had a rare ice-free rocky harbour that would provide a stable foundation for the station buildings and access to a glacier for fresh water. A vessel could anchor in deep water within 60 metres of the shore, while the expeditioners would have relatively easy access to a variety of different landforms around the base as well as the plateau beyond it. 'I hoped,' Law wrote, 'that, with a toehold on the Antarctic continent, I could persuade the government to expand the bridgehead in 1955 and gradually build up a full program of research.'[8] Law had planned that a party of ten men, led by Robert Dovers (whose father had served as Mawson's cartographer in 1911–1912), would construct the station and remain there for the year, but the early signs were not encouraging.[9] One of the Weasels—small motorised oversnow vehicles running on tracks—broke through the thin sea ice before it could land, and Dovers' party spent an uncomfortable night in the other Weasel, which had to be moved every few hours to ensure that it did not meet the same fate.

← A 1939 map of Antarctica showing the coastal regions of the Australian Antarctic Territory, and an inset depicting the continent in relation to the principal landmasses in the Southern Hemisphere.

When Dovers finally reached Horseshoe Harbour, he reported that the wind was worse there and proposed that his party spend the year on a small offshore island and try to find a better site for the permanent station during the following summer. Law thought the island a 'dismal place', with half of it occupied by an Adélie penguin (*Pygoscelis adeliae*) colony and the other half covered in boulders and coarse granite sand. With the ship beset by ice and storms tearing at the moorings of the barges, he realised that the conditions had lowered morale among Dovers' team and insisted that the station be established on Horseshoe Harbour: '[M]orale, as I had learned at Heard Island over several years, is largely conditioned by weather.'[10] The expeditioners spent the next week unloading stores and equipment for Australia's first base on the Antarctic continent, then the *Kista Dan* departed, leaving the ANARE team alone at the new base. The radio operator and second in charge, Louis (Lem) Macy, wrote in his diary:

> *We all piled into the 'Rodger Vincent' and drove alongside the ship and yelled insults etc to each other and away went the good old 'Kista'. We returned to camp and John and Bob D entered the kitchen to be greeted with a terrific explosion caused by the tin of King Sound salmon blowing up in the oven where Geoff had put 4 or 5 tins to thaw out before going to the ship—a fearful mess in the oven and the remaining tins were practically round with pressure.*[11]

As it happened, Dovers' team was extraordinarily productive during that first year. They completed the station buildings and undertook several traverses into the plateau to map the Framnes Mountains, carry out geological studies of rocky outcrops, collect samples of plant and animal life, perform meteorological observations, and establish radio communications with Australia. One party surveyed a safe route for dog sleds and Weasels to pass through a heavily crevassed area between Mawson and Mount Henderson. Another discovered a large emperor penguin (*Aptenodytes forsteri*) colony near the Taylor Glacier, which was given early protection as an Antarctic Specially Protected Area (ASPA) in 1959. Dovers led a group across sea ice to inspect the Scullin Monolith, a prominent rock first sighted by Mawson from the air in 1930 during the British, Australian, New Zealand Antarctic Expeditions (BANZARE) and named for Prime Minister James H. Scullin after Mawson landed on the rock in 1931.[12]

↑ Ice being hacked from a small glacier behind Mawson station and stacked outside the kitchen door for drinking water.

↗ Establishing post offices in Australia's remote outposts not only helped to connect expeditioners with their loved ones at home. They also served to demonstrate Australia's territorial claim in the region.

→ Two of the huts originally deployed on Heard Island were relocated to the new Mawson research station in 1955.

Voyaging to Antarctica

The voyage across the Southern Ocean has been an integral part of the Antarctic experience for generations of Australian expeditioners. Since members of the Australasian Antarctic Expedition (AAE) sailed to Antarctica on the wooden steam yacht *Aurora* in 1911, ships have provided a crucial lifeline for the existence of Australia's remote outposts in Antarctica and the subantarctic islands, carrying essential personnel, supplies and equipment, as well as bringing expeditioners home after long months of isolation.

In 1953, the newly formed ANARE chartered the *Kista Dan* (1953–1957), the first of a succession of polar icebreakers made by the Danish J. Lauritzen Line. *Kista Dan* was succeeded by *Thala Dan* and *Magga Dan* (1957–1962), the latter ship allowing ANARE scientists and surveyors to explore the entire coastline of the Australian Antarctic Territory. The *Magga Dan* was followed by the much-loved *Nella Dan* (1962–1987), named after the Australian artist Nel Law (the first Australian woman to set foot on Antarctica, she was the wife of Dr Phillip Law, ANARE leader and Director of Australia's Antarctic Division). *Nella Dan* navigated the notoriously treacherous winds and currents of the Southern Ocean between Hobart and Australia's Antarctic and subantarctic research stations for 26 years and provided the platform for ice reconnaissance and marine science in some of the most treacherous waters on Earth. The *Nella Dan*'s illustrious career on the Southern Ocean came to an abrupt end in 1987 when the ship was blown ashore during a routine refuelling stop at Macquarie Island. There were few dry eyes among ANARE personnel when the decision was made to scuttle the *Nella Dan* in deep water off the island, inspiring an outpouring of affection in poetry and song to honour the 'little, plucky red ship'.[13]

During the 1980s, while Australia's major rebuilding program was underway at the Antarctic research stations, *Icebird* (later *Polar Bird*) was used to transport equipment and supplies. The first Australian-made icebreaker, *Aurora Australis*, came next, serving as Australia's Antarctic flagship for the next 30 years (1989–2020). Affectionately known as the 'Orange Roughy' (a reference to bright reddish-orange deepwater fish inhabiting southern Australian waters), the *Aurora Australis* was a familiar sight docked at Princes Wharf in Hobart over the winter months. During its career, the ship undertook 150 Australian research and supply voyages and carried some 14,000 expeditioners between Australia and Antarctica. Meanwhile the crew and expeditioners experienced everything that modern Antarctic voyaging has to offer, from fires and hurricane-force winds to becoming beset by sea ice.[14] Icebreakers still provide a crucial lifeline for Australia's Antarctic and subantarctic stations and a base for scientists undertaking ocean research. In October 2021, Australia's newest research and resupply vessel, *Nuyina* (pronounced 'noy-yee-nah' and meaning 'southern lights' in palawa kani) arrived in Hobart. It marked the start of new era of Antarctic voyaging, 110 years after the *Aurora*—Mawson's three-masted, reinforced wooden sailing ship—left Hobart for Antarctica.[15]

↓ The first Antarctic voyage of Australia's new icebreaker *Nuyina* in 2021–2022 marked the beginning of a new era of ship-based research and resupply voyages to Australia's Antarctic and subantarctic research stations.

The Vestfold Hills

In 1957, three years after establishing Mawson station, Phillip Law again stood on the deck of the *Kista Dan* as he scanned the rocky, ice-free coastline of Prydz Bay in the Vestfold Hills. This time he was looking for a suitable site on which to establish Australia's second permanent station as part of the International Geophysical Year (IGY).

The Vestfold Hills had intrigued Mawson when he first saw them in February 1931 during the second BANZARE. The Norwegian whaler Captain Klarius Mikkelsen, who landed his vessel there in 1935, thought that the terrain reminded him of hills in the Norwegian province of Vestfold, and he named it accordingly. Sir Hubert Wilkins, a South Australian aviator and Arctic explorer, visited the Vestfold Hills in January 1939 as pilot and observer for the American explorer Lincoln Ellsworth.[16] Wilkins made two landings—one on the Rauer Islands and the other on a rocky outcrop in the Vestfold Hills—in an effort to thwart Ellsworth's plan to claim the area for the United States. At each landing site, he proclaimed Australia's right to 'those parts of His Majesty's dominions in the Antarctic Seas'. At the rocky outcrop, he planted an Australian flag and deposited a makeshift cylinder made of two enamel coffee jugs. It contained the handwritten proclamation wrapped in a copy of the Australian geographical magazine *Walkabout*. The site became known as Walkabout Rocks.[17]

The Vestfold Hills is the largest ice-free oasis in Antarctica and one of several in the Australian sector. The valleys contain more than 300 lakes, some hypersaline, others completely fresh. The area around the lakes is covered by yellowish sediments, studded with millions of stones. Law thought that the area looked more like a gibber desert in central Australia than a part of the frozen continent.[18] He wrote:

> *The average height of the hills is about 400 feet [120 metres] above sea level and the general scenery, with the warm chocolate-brown rock, the blue and white semi-frozen fjords, and the paler blue-green ice-free lakes, is glorious.*[19]

Offshore lies a scattering of small islands that provide sanctuary for several million Adélie penguins during breeding season.

Law searched for a suitable site for the station for three days as the *Kista Dan* cruised up and down the coast and an aircraft and small boats scouted for potential landing spots. Finally, on 12 January, he decided on a pebbly terrace rising above a small, sandy beach. There was no local source of water, snow or ice, but the ship's engineer built an electric distilling plant to use with sea water. Using Army DUKWs (six-wheeled amphibious trucks, known as 'ducks'), the expeditioners unloaded equipment and supplies for the four-man team, led by ANARE meteorologist Robert Dingle, who would remain for the year to establish the station and undertake scientific work. The next day, Law officially named the new station 'Davis' after Captain John King Davis, the celebrated captain of Mawson's expedition ships *Aurora* (1911–1914) and *Discovery* (1929–1930). For the next week, the expeditioners built a sleeping hut, community hut, engine hut, store hut, balloon-filling shed and auroral observatory, as well as undertaking photographic flights over the area and nearby Amery Ice Shelf.[20] Australia, Law announced on his return to Melbourne, was ready to participate in the IGY.

↓ A small freshwater lake in the heart of the Vestfold Hills, the largest ice-free oasis in Antarctica.

→ An exploration party with their two DUKW vehicles on a beach at the foot of a steep escarpment of the Vestfold Hills in 1955.

Australia's Antarctic and subantarctic permanent research stations

Mawson (67°36'10" S, 62°52'23" E)

Mawson, established in 1954, was Australia's first permanent research station in Antarctica and is the longest continuously operating station on the continent. It is the smallest and most westerly of Australia's three continental research stations, and the most distant from Australia (5,200 kilometres southwest of Perth), taking around 10 to 12 days to reach by ship from Hobart.[21] Mawson can accommodate up to 60 people in summer and 20 in winter. It sits on an outcrop of exposed rock overlooking the harbour, protected from the swells of the Southern Ocean, although rather less immune to the blizzards that blow down from the continental plateau just 400 metres away. Importantly for a station ever reliant on the annual resupply mission, ships can anchor within 100 metres of the station.[22]

Davis (68°34'35" S, 77°58'08" E)

Davis, 4,736 kilometres south of Perth and the most southerly Australian research station on the continent, lies on the edge of the Vestfold Hills on the eastern side of Prydz Bay in Princess Elizabeth Land. It was established in 1957 as part of Australia's scientific program during the International Geophysical Year (IGY) and is named in honour of Captain John King Davis, captain of the *Aurora* on Mawson's AAE and other Antarctic voyages. Davis is located within a triangle of ice and ocean: the Sørsdal Glacier lies to the south,

the continental plateau to the east, and the sea to the northwest. Today the station can accommodate up to 120 expeditioners over summer and about 20 over winter and, as with Mawson and Casey, features a large dining room and modern facilities.[23]

Casey (66°16'55" S, 110°31'39" E)

Casey was the last of Australia's three permanent continental research stations to be completed (in 1988). It is the closest station

to the Australian mainland (3,880 kilometres due south of Perth), located in the Windmill Islands just outside the Antarctic Circle, on the edge of the Antarctic ice sheet. Casey was built to accommodate up to 160 expeditioners in summer and around 16 to 20 in winter, with nearby Wilkins Runway enabling personnel to reach Casey by air. Air travel has dramatically reduced the voyage time; it now takes less than five hours, compared with ten days by ship. Casey's main accommodation building, known as the 'Red Shed', boasts a hydroponic garden, supplying fresh vegetables to supplement the diet of winterers. The island location of the station means that the area abounds with Adélie penguins, giant petrels, skuas and snow petrels. In 1993, Casey became home to the Antarctic Meteorological Centre, providing satellite imagery to the World Meteorological Centre in Melbourne. In 2008, the station also became the base for the International Collaboration for Exploration of the Cryosphere through Aerogeophysical Profiling (ICECAP) project, designed to study the glaciology and bedrock of the East Antarctic Ice Sheet. Seven years later, it hosted the world's first underwater ocean

← ANARE meteorology weather forecaster Neville Martin releases the evening weather balloon which collects data of local atmospheric and weather conditions around Davis station, 1997.

acidification experiment in Antarctica, which involved chambers being lowered beneath the sea ice to observe the effects of variations in carbon dioxide levels on seafloor life.[24]

Macquarie Island (54° 37'12" S, 158° 51'40" E)

Macquarie Island first became the focus of scientific research in 1911–1914, when members of Mawson's AAE carried out extensive fieldwork on and around the island, as well as mapping it, recording meteorological data and collecting specimens. The Commonwealth Meteorological Service assumed responsibility for meteorological observations on the island from 1913, but the service was discontinued after the loss of an expedition member and the disappearance of the relief ship *Endeavour* in 1914. Mawson visited the island again in 1929 and 1931 during the BANZARE, and in 1947–1948 ANARE established a research station at the northern end of the island, where it remains today. Macquarie Island, or 'Macca', is occupied year-round by some 40 expeditioners over summer and 16 to 20 over winter. The island measures 34 kilometres long and five kilometres wide and, together with its surrounding waters extending out 12 nautical miles, was listed as a World Heritage Area in 1997. The reserve is managed by the Tasmanian Parks and Wildlife Service of the Tasmanian government's Department of Primary Industries, Parks, Water and Environment. It lies 1,542 kilometres southeast of Hobart in the path of westerly winds known as the 'Furious Fifties'. Research is conducted by the Australian Antarctic Division in cooperation with Tasmania.[25]

The International Geophysical Year

In the years after the Second World War, a small group of prominent scientists put forward a bold proposal to undertake a systematic worldwide study of Earth and its planetary environment, timed to coincide with the period of maximum sunspot activity. The proposal led to the IGY, conducted between 1 July 1957 and 31 December 1958, with scientists from 67 countries undertaking research in a range of disciplines, including meteorology, geophysics, glaciology, oceanography and seismology. Antarctica was still an enigma, the largest unexplored continent on the planet, and the IGY offered the perfect opportunity for nations to put aside their contest for territory in pursuit of a coordinated scientific program that would yield benefits for the whole world. As Phillip Law observed:

> *Man's knowledge of the Antarctic continent before 1954 was fragmentary. There were many stretches of coast that had never been explored and many mountains that had never been sighted. The Antarctic plateau was almost wholly unknown.*[26]

The IGY was a phenomenal success, revolutionising scientific understanding of Earth. Participating nations established some 60 stations in Antarctica, including a US station at the South Pole and a Soviet station at the Pole of Inaccessibility, the point most distant from the coastline. For the first time, scientists could examine the ice sheet, from its

mountain peaks to its bedrock, using rocket and satellite technologies and the latest instruments, but the IGY's legacy went far beyond the collection of data. It transformed Antarctic geopolitics by laying the groundwork for the international Antarctic Treaty, signed in Washington on 1 December 1959 by the 12 nations with scientists active in Antarctica during the IGY.[27] The signatories, including Australia, agreed to put aside their territorial claims, ban nuclear explosions and nuclear waste disposal, and dedicate the continent to peace and science. It was a remarkable

← The International Geophysical Year revolutionised scientific understanding of planet Earth

demonstration of international diplomacy in the midst of Cold War tensions, and it would profoundly influence the nature of human activities in Antarctica.[28]

In 2007, 50 years after the IGY, scientists from 63 nations came together in another ambitious collaborative scientific program known as the International Polar Year (IPY). The IPY aimed to expand scientific understanding of the polar regions and monitor the impacts of global warming on these sensitive frozen environments. As with the IGY, Australian scientists contributed their expertise by leading several major projects, including a census of Antarctic marine life in the waters off East Antarctica, and a study of climate in Antarctica and the Southern Ocean.[29]

From Wilkes to Casey, the long grey caterpillar

As the IGY drew to a close, the United States was left with a station that it no longer required, and the American government decided to offer it to ANARE on the basis that it would be co-administered by the two nations. The arrangement offered considerable advantages to Australia's Antarctic program: Wilkes station was small but exceedingly well equipped. It was closer to the Australian continent than the other two stations and conveniently located for recording Antarctic weather patterns that influenced southern Australia's climate. It was also strategically positioned close to the south magnetic pole, as well as being near two sites of scientific interest: an expanse of coastal moss beds and a prominent ice dome that Australia subsequently named the Law Dome for Phillip Law, in honour of his contribution to Australian Antarctic exploration and science.

Wilkes proved popular with the Australians. It was more spacious, warm and comfortable than either Mawson or Davis, and the recreation room had a pool table and a table-tennis table, a library and an extensive range of classic Hollywood films. The Americans, it seemed, had not wanted for anything during their brief residency. The United States allowed Australian expeditioners to use any of the stores or facilities that remained at the station, providing they removed nothing and reported annually on consumption of the supplies. It soon became clear, however, that Wilkes had its problems. It was situated in a depression, and the buildings were especially prone to heavy snowdrifts that, over time, threatened to engulf the entire site. To make matters worse, it had been built by a United States Navy crew in just 16 days, and the temporary structures made of plywood panelling were inclined to rot as the ice melted. Fuel leaks added to the dangers, having seeped under the buildings and soaked into the timbers. As glaciologist Vin Morgan later observed:

> Wilkes is basically buried under snow, so it was a real warren in winter and pretty dangerous because it all had kerosene and oil all over the place and there was very little way out if there'd been a fire or anything in winter.[30]

Within a decade, the Australian government decided to build a replacement station on nearby Vincennes Bay, just outside the Antarctic Circle. The replacement station (Repstat) was officially named Casey research station after Australia's governor-general Richard Casey. As Minister for External Affairs, Richard

→ An aerial view of Wilkes station in 1962.

↓ The formal handover ceremony of Wilkes station to Australia in February 1959.

Casey had worked closely with Law to develop Australia's ANARE program during its formative years in the 1950s and 1960s, including supporting the establishment of Mawson and Davis stations. Casey was constructed on the opposite side of the Bailey Peninsula, and it was as different from Wilkes as it could be.

The new station comprised a series of separate buildings for different functions, including a recreation room, mess, doctor's surgery, science buildings and a darkroom. The snowdrift that had enveloped Wilkes was still fresh in the minds of the designers when they devised a walkway linking the dozen or so box-like buildings. The walkway was protected on the windward side by sheets of galvanised iron and the whole tunnel-like structure, fondly known as 'the long grey caterpillar', was elevated above the ground using scaffold piping so that snowdrift could simply blow underneath the buildings rather than pile up around them.[31] Moreover, it was constructed over four years by ANARE expeditioners based at Wilkes station. The tradesmen would visit the new site for a week at a time to complete the more complex tasks, and everyone else would lend a hand during their time off. Progress was slow, however, and after three years the Australian Antarctic Division appointed a small team of six men to complete the station. Plumber Rod Mackenzie, who was one of the six appointed to the 'Repstat group', recalled how they would work from 8.30am to 6.30pm, six days a week, throughout the dark winter months. They had to be self-sufficient out on the building site and, as they worked, they listened to music and

weather forecasts broadcast from the radio station at Wilkes. Mackenzie, who had trained in breadmaking, would bake 32 pounds (14.5 kilograms) of bread each week for the group.[32]

With the completion of the first Casey station in 1969, the Antarctic Division closed Wilkes, although Australian expeditioners continued to visit to 'borrow' spare parts and stores and enjoy the occasional 'jolly' (excursion) away from the main station.[33] Keith Gooley recalls visiting Wilkes during his year at Casey in 1974:

← Mike Stracey exploring one of the abandoned buildings at Wilkes station in 1974.

→ Australian expeditioners could still find unopened tins of jam in their original boxes in 1997, nearly 30 years after the Antarctic Division closed Wilkes station.

The station had just been abandoned, left [with] breakfast cereal on the table and the like. It was filling up with ice about a metre deep on the floor, and you had to duck your head in places.[34]

Many of those who experienced life at Casey considered it superior to other stations. Casey was prone to the same extremes of weather that had plagued Wilkes, but the distinctive elevated structure resisted the build-up of snow and ice. Law thought it a brilliant idea, although the strong winds blowing down from the plateau tended to howl through the scaffold tubing and guy wires and make the entire structure vibrate.[35] A blizzard would stop any outside work altogether, unless an emergency job beckoned. Charlie Weir recalled on one occasion that the winds reached 200 knots (370 kilometres per hour). When the anemometer was blown away and the protective iron sheeting on the tunnel started to lift, it seemed to him that the whole station would disintegrate:

Where do we go? There's no way you could go outside in a wind that strong. And that's the only time I was a bit frightened, I didn't sleep that night.[36]

Mercifully, the station stayed put, although the radio aerials did blow away and it would be a week or more before communications were restored with the outside world.

When the Australian journalist and writer Stephen Murray-Smith visited Casey aboard the *Icebird* in 1985, he was confronted by a bleak vision of grey metal and mud. The station design, 'either as a visual spectacle or a living experience, defies Heath Robinson's most bizarre dreams', he declared, while the inhabitants could spend an entire season in this 'enclosed, interconnected little world' and rarely have to venture outside.[37] (Heath Robinson was a British artist whose name became synonymous with absurdly complex inventions with no practical use.) By then, Old Casey had been accommodating ANARE personnel for nearly 20 years, although its design, he noted, had not been repeated elsewhere on the continent. Nevertheless, the distinctive tunnel arrangement had its advantages. Inhabitants could move between the main buildings and their dongas (or sleeping quarters) in relative comfort, albeit having to negotiate ice on the floor in the unheated tunnel. It offered opportunities for more social interaction among the expeditioners, and most of the Casey 'tunnel rats', as they were known, regarded the old station with affection. As station leader Graeme Manning recalled, in the 14 months he was stationed at Casey he saw every person at least once a day either in the tunnel or the mess room.[38] This insular existence also had its critics. As one expeditioner put it, they were less inclined to venture outside 'to feel Antarctica, to appreciate Antarctica, to be one with the whole thing'.[39]

Over the years, the tunnel proved difficult to maintain and the Australian government decided to replace it as part of a major 1980s rebuilding program across all three continental stations. The tunnel was finally decommissioned in the early 1990s, and its materials shipped back to Australia. Unlike the low-cost tunnel complex built by expeditioners in their spare time, 'new Casey' was designed by the Australian Department of Administrative Services and built using professional tradesmen recruited from the Australian building industry. The design team had experimented with different materials and methods of construction. As a result of this process, Australian Construction Services developed the Australian Antarctic Building System (AANBUS), in which buildings were finished and fitted out in Australia before being shipped to Antarctica for construction. The new approach revolutionised Antarctic building design and the model was subsequently adopted by other countries.[40]

New Casey would have steel-framed buildings anchored into concrete foundations, and the external walls would be clad with thick steel panels filled with an insulating core of polystyrene foam for greater safety. There would be no tunnel connecting the station buildings. Instead, the buildings were set apart and painted in bright colours intended to bring some visual relief to the stark, monochromatic landscape. The Red Sheds, as the main buildings were called, would contain a large common room, a theatre, library, medical centre, mess rooms and kitchens, as well as individual bedrooms. The new, larger buildings were set above the ground, in the manner of the old station,

↑ The distinctive tunnel building at the Repstat (Casey station) where the resident expeditioners were fondly known as 'tunnel rats'.

↠ New Casey in 2005 with its colourful buildings, including the main living spaces known as the Red Sheds.

to minimise the build-up of snowdrift, and their long sides were aligned so as to be parallel to the prevailing winds.

Some thought the scale and cost of the new station difficult to justify given the small number of expeditioners who would use it. Others expressed concern that the attention to creating more space and privacy would discourage the kind of close social interaction that had been a hallmark of the older stations.[41] Nevertheless, the sheer scale of the new station, and its prominent location at the top of the ridge above Vincennes Bay, seemed to underscore the fact that Australians were here to stay. As Murray-Smith walked between the old and new stations in 1985, he contemplated how Casey seemed to symbolise Australia's new relationship with Antarctica: 'the crossing of a divide between the old way of doing things in Antarctica and the new'. He wondered what Douglas Mawson would have made of it all.[42]

A congregation of emperor penguins is an island fortress
of essential life, maintained successfully in time, amidst
the most fearsome assaults of darkness, cold, and turmoil
that the inanimate world can deliver. It will survive the
course of all men's expeditions.

JOHN BÉCHERVAISE, 1963[1]

Station

← Emperor penguin
huddle, 2013.

t was April Fool's Day 1959, and there were ominous signs of things to come. An unusually high ocean swell was making the sea ice in the harbour at Mawson station levitate alarmingly. The next day was clear enough, but violent 'willies' (winds) were picking up loose materials and flinging them around. 'All day, out at sea, the winds fumed in wild circles,' wrote the Australian polar explorer and writer John Béchervaise. 'They tore up ice six inches [15 centimetres] thick in sheets, and sent columns of spray skyward.'[2] The equipment measured strange changes in air pressure, and the day ended with a flame-like sunset. Everyone was uneasy. It had been nearly 50 years since the first Australian-led expedition experienced winter conditions on the continent, but Béchervaise's experiences would show just how fraught living and working in this extreme environment continued to be.[3]

Béchervaise, an experienced mountaineer, had earned a reputation as a respected leader from his time as Officer-in-Charge (OIC) on Heard Island in 1953. Glaciologist Peter Keage remembered him as an imposing character:

> In his prime he was big in stature, deep in voice and blessed with presence—you knew when 'Beche' entered the room. He was outward-looking, engaging, a measured risk-taker, a good writer, artist, photographer and rich in that intangible and rare quality—leadership.[4]

By the time Béchervaise arrived at Mawson, the station had grown into a village with some 40 living huts. There was also a central area with a surgery, power station, store huts and administration offices, and an array of field huts with equipment for monitoring and gathering data on meteorology, cosmic rays, geomagnetism, seismology, the aurora, the ionosphere and biology. Close to shore, at the edge of the ice, a hangar housed the station's two light aircraft.[5]

As winter approached, the ANARE expeditioners were racing to finish setting up equipment before the fickle Antarctic weather brought an abrupt end to their fieldwork. Béchervaise recorded the events that followed:

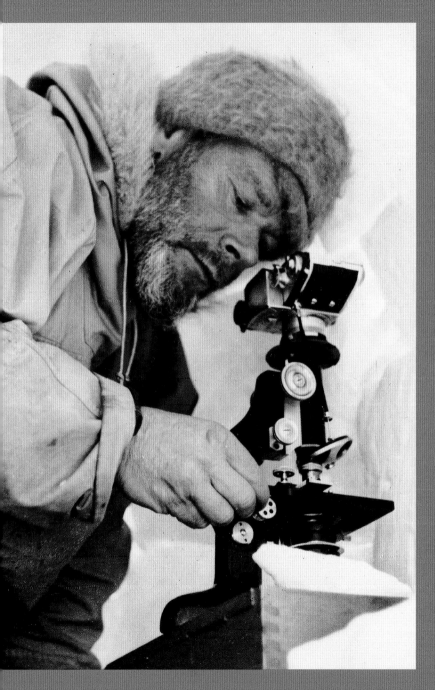

On the hill, Norris works successfully below his auroral domes, Van Hulssen slowly gains on his recalcitrant ionosonde [ionospheric radar], and continues to record the movements of meteor trails, and Dunlop successfully counts his cosmic particles. Widdows..., Onley, and Peake-Jones have settled to met. routine. Cosgrove, in addition to being heavily involved in a programme of experimental teletype installations, has worked a twenty-four-hour shift on the radiosonde receiver; he has also set up our new record-player in 'Weddell' ... Peake-Jones, as fire officer, is working in all spare moments towards our maximum safety from the station's worst danger.[6]

← John Béchervaise, teacher, author, poet, explorer and photographer, served as OIC on Heard Island in 1953 and at Mawson in 1955 and 1959.

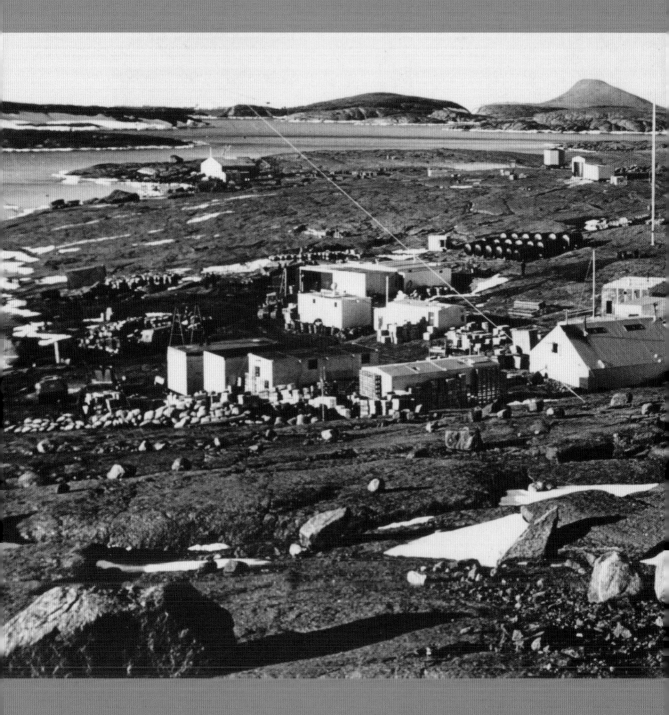

↑ An early photograph of Mawson research station in summer, taken by Phillip Law, who oversaw the establishment of Mawson, Casey and Davis stations and served as Director of Australia's Antarctic Division from 1949 to 1966.

The next day, fire consumed the newly constructed powerhouse. The expeditioners fought the flames to the point of exhaustion, but the 'great fire', as Béchervaise called it, engulfed the structure and pitched the station into darkness. 'We all stood on the bitter edge of the world,' he wrote.[7] Indeed, the fire would be an ominous portent of things to come. The station was in the midst of the windiest and most 'blizzardly' autumn in its brief history. The temperature plummeted, reaching as low as −45°F (−43°C), and it was only by a stroke of luck that the men had been able to salvage the big generator from the fire and build a new powerhouse in record time. The scientific program was soon resumed, and the station was a hive of activity once more. Apart from research and construction work, there were weekly French classes, Sunday night talks, skating and skiing, games, music and the all-important 'WYSSAs' (radiograms from loved ones). The team even marked the Queen's Birthday with a flypast conducted by the station's two Beaver aircraft, and held a dawn service to commemorate Anzac Day.[8]

The monster is fully aroused

Then came the worst blizzard in living memory. For Béchervaise, that night in July was 'the wildest, most forlorn night I have ever known'. It even surpassed the storm that he and two companions had weathered on Heard Island when they been virtually imprisoned in their tent for five days. Béchervaise recalled how they had been kept busy shovelling snow away from the tent:

> *We consolidated our space by pressing our backs against the tent fabric and the mounting snow, which eventually met over our heads; beneath this vault we could then remain safe through just such a night as this.*[9]

Yet, within his tiny room at Mawson, Béchervaise felt even less secure than in that tent. The Mawson accommodation comprised a string of huts with space between them so that any fire that might break out in one could not spread to the rest. Each hut was secured by steel cables set into the exposed rock, but now this blizzard was trying to tear them from their moorings. Béchervaise wondered whether the whole station might be blown away. In such conditions, everyone was afraid.[10] Despite advances in

↑ **Blizzards come with the territory in Antarctica. Frank Hurley's photograph of a blizzard at Cape Denison during the AAE in 1913 captures perfectly the extreme conditions experienced by expeditioners since Mawson's day.**

transport and communications technologies emerging from the Second World War, and the introduction of prefabricated buildings designed to withstand Antarctic conditions, expeditioners still had to deal with the physical risks of living and working in the Antarctic region. At Mawson, anyone leaving their hut in poor weather was required to follow a procedure: before setting out, they would advise the others of their intended destination by radio, then feel their way along a specially installed 'blizz line' in the darkness. It was crucial that they communicate their safe arrival to the others. Béchervaise wrote:

> *There is a sort of communal consciousness; everyone is aware of the meaning of the bells. 'That'll be Ross, arriving at cosray.' 'Now, why hasn't old Chiefy rung back from the hangar?' … To lose the line between two huts in this weather could be a final error, unless one were sufficiently dressed, and possessed the fortitude to sit huddled against the blizzard like a dog for the hours or days it might last.*[11]

Daylight brought no relief. Visibility shrank to the space between eyes and boots, and the blizzard seemed to metamorphose into a living thing. 'The monster is fully aroused,' wrote Béchervaise, 'and quite contemptuous of human resolution ... Often it seems that we are fighting a cunning, sentient foe, striving always to outwit us.'[12] As the men huddled in their huts, they took turns to stoke their fires so that the dense particles of wind-driven snow that melted in the warm flues and air vents could not be allowed to freeze over again. To do so would mean the build-up of carbon monoxide within the hut, and certain death to its occupants.[13] Two of the men, David Norris and Les Onley, were forced to endure the blizzard in an auroral observatory located at Taylor Glacier some 80 kilometres to the west of the station. For them, matters took a turn for the worse. After initially making radio contact indicating that the blizzard was not as severe as at Mawson, the station lost contact with them. The weather afforded limited options to reach the men but, after several days of de-icing one of the Beaver planes, Béchervaise and two others made the westward journey by air.[14] As they circled to find a landing spot, they realised that the observatory was no longer there and all that remained was a black smudge staining the snow. The building had been destroyed by fire, and its two inhabitants had escaped, taking shelter in a stores hut. It could have been worse, of course, but for Béchervaise as OIC it was yet another blow for this ANARE.

The Antarctic winds had not finished with them yet. Another hurricane in December of that year struck the icefield airstrips above Mawson, suddenly veering towards the station and ripping the two Beaver aircraft from their cable moorings. While some of the team tried to save them, others battled the storm to reach the sledge caravan, which had slumped into the ice and was threatening to escape its five-tonne steel guys.

> *Thereafter, for seventeen depressing and alarming hours, until it was possible to drive a weasel down to base, men kept watch in case the sledge caravan should break loose. Men slept in their windproofs, some with c[r]amponed feet.*[15]

They watched as the second aircraft finally yielded to the storm:

> *The wing suddenly came away and seemed to be hurled directly at the caravan. It passed a few yards to our westward and, above the scream of the wind, we heard it momentarily crash to the ice before being borne away ... 'I reckon,' said Mac with some relief, 'that I'll make a nice pot o' tea!'.*[16]

↑ De Havilland DHC-2 Beaver aircraft and DUKW military trucks
on board the *Thala Dan*, near Wilkes Land, Antarctica, 1962.

The team at Mawson felt the loss of the two aircraft even more than all the other misfortunes of that year. The planes had been indispensable for mapping the features of the plateau and undertaking fieldwork, including a traverse called Operation Icefield and Operation Drift, which involved radiation work and microphotography of drift particles.

During that winter, Béchervaise often contemplated the peculiar sense of 'Antarctic time' that pervaded life in this region. A 'ponderous granite erratic' that had broken in two and taken up residence at Mawson remained there, 'scarcely time-marked',

while 'this silver scientific village' would soon be gone.[17] Then, with the arrival of the relief ship, 'a new kind of time arrives … Men are all wildly busy, yet they don't really achieve any more for their hours; the days are suddenly brief and fleeting, as in cities'.[18] As he watched the rituals of a great emperor penguin colony near the Taylor Glacier, where the ANARE observatory had been reduced to ashes, he pondered how these rituals had existed long before Antarctica was even imagined by human minds. He was equivocal about his time at Mawson:

> *Personally, a man carries almost nothing objective away from Antarctica. His are the truly intrinsic values of memory, the only true stuff of personal consciousness, which, nevertheless, may be no unalloyed blessing.*[19]

The 'dome of Antarctica', he observed, is spread about the South Pole 'like white clay yet to be centred on a gigantic potter's wheel, spinning steadily … on the axis of the earth'. He thought it 'the most frigid region that man will ever know until he reaches night on the moon' and wondered why Australia had chosen to occupy a place that seemed 'part of a dying world—pallid, remote, clouded, and formidable, surrounded by dark, ice-flecked water swirling with vast cyclones advancing slowly round the earth'.[20]

The 1959–1960 ANARE team at Mawson had survived a near shipwreck (when their vessel struck a submerged rock), two fires and severe blizzards. Nevertheless, even as they prepared to leave Mawson after the tumultuous year, Béchervaise was still reluctant to depart, marvelling at the colours and light of an Antarctic summer: 'It is brief and capricious', he wrote, 'but every perfect hour possesses a quality of eternity, a prolongation of the moment'.[21]

← A true-colour mosaic of Antarctica produced from NASA's 2013 *Terra* satellite images.

Calling Antarctica

Antarctica's extreme isolation poses particular challenges for those who venture there, and the ability to communicate with home has always been one of the most important aspects of life on Australia's research stations. Indeed, prior to the introduction of satellite technology, infrequent and unreliable communication was a major factor contributing to poor morale among expeditioners, and could also be distressing for their families at home, particularly for those affected by illness or other personal difficulties.

The world's first radio communication to and from Antarctica began during Douglas Mawson's Australasian Antarctic Expedition (AAE) in 1911–1914. Mawson's plan was to use a wireless transmitter to communicate with Hobart via a relay station on Macquarie Island, although early attempts to transmit messages were hampered by weather conditions and Mawson initially considered the wireless to be one of the expedition's biggest failures. By early 1913, however, Sidney Jeffryes was able to alert the *Aurora* to Mawson's return from the tragic sledging journey that had claimed the life of Xavier Mertz and Belgrave Ninnis. With the weather deteriorating, Captain Davis decided to rescue the party stranded at Western Base rather than to return to Commonwealth Bay, leaving Mawson, Jeffryes and the five surviving expeditioners to see out another winter at Commonwealth Bay. As winter approached, Jeffryes made Antarctic history by establishing regular radio communications with Hobart, and the men were able to receive weather reports and news from the outside world.[22]

Even with the establishment of permanent stations, radio communications were still rudimentary, with expeditioners relying on sending and receiving messages in Morse code at scheduled times. Much of this early radio contact occurred between stations to exchange weather information. Otherwise, expeditioners used amateur 'ham' radios for contact with friends and family at home. Rod Mackenzie recalled feeling cut off from the rest of the world in the early days of ANARE operations: 'I guess it's the lack of communication, the fact that you can't get in or out, people can't get down. If anything goes wrong you're stuck there'.[23] The Australian Antarctic Division introduced a whole

→ Norma Ferris presenting the ABC's radio program *Calling Antarctica* in 1969. The program ran for nearly 40 years, bringing news, family greetings and music requests to those stationed at Australia's remote Antarctic research stations.

new vocabulary for Antarctic communications with a system of five-letter codes, and employed a welfare officer whose role included transmitting coded letters by cable between ANARE expeditioners and their families. 'WYSSA', for example, translated to 'all my (our) love darling'.[24] When the ABC's Radio Australia began broadcasting a weekly half-hour radio program, *Calling Antarctica,* in 1948, relatives of expeditioners stationed at Casey, Mawson, Davis and Macquarie Island could visit the ABC studio in their nearest city to record messages or request to have their letters broadcast over the air.[25] The program used only female presenters, and the men could listen to current affairs and music requests from friends and family. Joan Saxton (wife of 1963 Wilkes OIC Richard Saxton) remembers how the men would stop whatever they were doing to listen in to those weekly broadcasts.[26] To help ease the problems of distance and isolation, the Antarctic Division's welfare officer, Mrs Rob (Shelagh Robinson), would also put expeditioners' families in touch with each other and even arrange to buy flowers or send a card for a special family event if expeditioners were going to miss it while in Antarctica.[27] *Calling Antarctica* ran for nearly 40 years, ending only in the mid-1980s, when satellites revolutionised communications between Australia and its Antarctic stations.

Midwinter madness

As Mawson's expeditioners discovered, wintering in Antarctica can have a strange effect on a person. For humans, the vast ice sheet is as alien as the surface of the moon or the depths of the surrounding ocean. In his book *Slicing the Silence,* Tom Griffiths reflected on the psychological effects of the long, dark winters on some expeditioners, and speculated about whether it was the extreme environment that had affected them or, conversely, whether Antarctica attracted certain personality types. Certainly, Australian expeditioners were not immune to what the British geologist and Antarctic explorer Raymond Priestley called 'polar madness'.[28] Living and working in confined spaces, particularly during the long, dark months of winter, with little opportunity for outdoor activities or time alone, invariably created psychological stresses.

Since Mawson's time, expedition leaders have recognised the importance of celebrating traditional events such as Christmas and birthdays as a way of boosting morale and creating bonds. One of the most widely anticipated celebrations in Antarctica is Midwinter's Day, when the winter solstice heralds the gradual return of the sun and longer hours of daylight. It commonly falls on 21 June but can occur on 22 or 23 June. On Australian stations, Midwinter's Day has traditionally involved a sumptuous banquet, extravagant attire and a theatrical performance, which, at least until the 1980s, involved a bawdy performance of *Cinderella* complete with a script modified for the occasion. Graeme Manning, OIC at Casey in 1979, recalls how preparations would begin three weeks before the day with wine-tastings to select the best for the banquet, menus prepared, photos selected to decorate souvenir menus, and rehearsals for the theatrical performance. Midwinter's Day at the old station would begin with tea or coffee delivered to each donga with a special midwinter edition of the station's newspaper, *Casey Chronic Truth*. That was followed by a champagne breakfast and

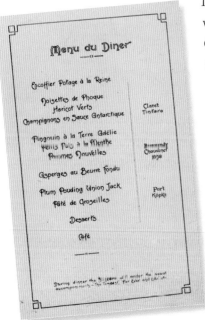

← The menu for the AAE's Midwinter's Day celebrations at Main Base, Cape Denison, in 1912.

→ The AAE Western Base Party celebrating Midwinter's Day at The Grottoes in 1912.

↓ Members of BANZARE celebrating Christmas on board the *Discovery* in 1930.

a Tunnel Run, with 25 people squeezing into the one metre-wide tunnel to race from one end to the other. The banquet lunch would be served around 1pm, followed by the performance of *Cinderella*.[29]

The first few weeks after Midwinter's Day were the most likely time for tensions to erupt among expeditioners. As Manning observed, the constant darkness could bring about 'big-eye', when people slept all day and found themselves unable to sleep at night. Routines were upended, and everyone would get 'touchy'.[30] Richard Penney, a biologist at Wilkes station in 1959 and 1960, referred to it as the 'grunt stage' of winter, when people were so familiar with each other that they would resort to a grunt rather than a greeting.[31] Such tensions would quickly dissipate with the first signs of spring. As plumber Rod Mackenzie remarked, 'One day you get a beautiful and magnificent blue sky and everyone goes mad, out throwing snowballs at each other and they get the bob-sleds out'.[32] They would rush out to welcome the return of penguins, taking bets on the first bird to be seen marching across the ice to its colony. The seasonal change had a profound impact on him:

> And it's so beautiful and so pure and pristine. So you never forget. It pulls you back. You want to keep going back. It takes you ten years to get it out of your system. I don't know if you ever do, really.[33]

← Expeditioners based at Australia's research stations have long enjoyed observing and photographing nearby penguin colonies during their time off.

← Australian geologist J.F. Ivanac makes friends with a rockhopper penguin on Heard Island in 1949.

With improvements in communications, transport and facilities from the 1950s, expeditioners gradually had more freedom to explore the station environs for themselves. Around Casey, for example, expeditioners would ski or abseil, go fishing, walking or driving during the longer daylight hours of summer. Alan White, who worked as the chef at Old Casey station in 1983, used to spend his free time after work out on a skidoo or trike, exploring the hills behind the station or visiting the penguin colony on Shirley Island in early spring to watch the adults returning to breed. He enjoyed taking photographs—a popular hobby among expeditioners—and would often go out walking late into the summer evening, then catch a few hours' sleep before getting up to work the next morning.[34]

During his time as OIC at Mawson in 1972, Neil Roberts wrote a yearbook documenting life at the station.[35] By then, Australia's first permanent station on the continent had acquired a distinctive atmosphere. The cramped 'Rec Room', though crowded and shabby, had served as the station lounge, bar, library, sports arena, debating hall, parade ground and boxing ring for 20 years. 'Yes we've loved that little room', he wrote, 'soaked in tradition and embalmed in the smoke, the stories, the laughter, the music and the new coat of paint of past years.' It was where the isolated expeditioners were able to gather and watch classic movies or enjoy the idiosyncratic interactions of station life.[36] Participating in outdoor games on the sea ice was also a big part of station life, at least when weather permitted. The station had its own Mawson Athletics Club, which staged races around West Bay to encourage its members to stay fit for sledging trips.[37] Nevertheless, preparations for undertaking scientific work took priority that year, including construction of the first underground laboratory for seismology and cosmic ray research. The hut was built over a 13 metre shaft excavated into solid rock, designed for monitoring cosmic ray 'events' occurring deep within it.[38] There were also various field trips to gather glaciological data from the ice sheet, sample water or establish fuel depots. One of the longer traverses that year involved delivering fuel by tractor train to Mount Cresswell in the Prince Charles Mountains, some 320 kilometres from Mawson. The fuel was to be used for a series of reconnaissance flights during the 1972–1973 summer ANARE, when scientists would undertake mapping, geophysical and geological surveys of the area. Invariably, such work was dictated by the weather and, at Mawson, the katabatic winds would arrive like clockwork at around 11am, bringing all outside work to a decisive halt.

On the frontier

Until the 1970s, Antarctic stations were almost exclusively male domains. The earliest women to voyage south had been wives, daughters and servants of naval men, sealers or whalers from the late eighteenth century. The first woman to visit Antarctica in a professional capacity was Maria Klenova, a marine geologist working for the Council for Antarctic Research of the USSR Academy of Sciences, in the summer of 1955–1956. Apart from four women scientists permitted to land on Macquarie Island in 1959, however, Australian women had to wait another two decades before official policies and cultural attitudes changed. There were rare unofficial visits. In 1960, for example, the Australian artist Nel Law was 'smuggled' aboard the ANARE resupply vessel by her husband Phillip Law as the ship prepared to depart from Perth bound for Mawson station. As Major Ian Toohill recalled, when the news broke that a woman was on the ship, there was an expectation that she would be asked to disembark; however, instead of provoking outrage, her presence seemed to arouse public interest in Australia's activities in Antarctica and delivered some much-needed support for the ANARE program. Nel Law had cemented her own place in Antarctic history as the first Australian woman in Antarctica and the first woman to set foot on an Australian continental station.[39]

In the summer of 1975–1976, Shelagh Robinson (welfare officer), Jutta Hösel (official photographer) and Elizabeth Chipman (information officer) from the Australian Antarctic Division arrived at Casey station, together with Dr Zoë Gardner, a British doctor recruited as medical officer for the Macquarie Island research station when the Division had difficulty finding a male medical officer willing to spend a year there. It was International Women's Year, and questions had been asked in the Australian Parliament about why there were still no women on Australian stations. The four women were welcomed courteously although some of the expeditioners were initially unhappy.[40] Despite lingering reservations, women began to make their presence felt. Jeannie Ledingham spent a full summer at Cape Denison in 1977–1978 and, in the following year, biologist Elizabeth Kerry became the first Australian woman to conduct scientific research on the continent.

Then in 1981, 70 years after Douglas Mawson's first Australian-led expedition, Dr Louise Holliday arrived at Davis, Australia's most southerly Antarctic base. She was the first woman to winter at an Australian research station in Antarctica, 20 years after Nel Law became the first woman to visit one.[41] Holliday had applied to serve as the station's medical officer after being inspired by stories told by her neighbour, Captain Morton Moyes, a naval officer who had served as a meteorologist with Frank Wild's Western Party during the AAE in 1911–1914. She recalled how Moyes, a family friend, would sometimes 'tell us stories of derring-do', such as the time he agreed to remain alone at the party's winter hut on the Shackleton Ice Shelf for three weeks while Wild led a sledging party. The group's return was delayed after losing a sledge, and Moyes would endure nine weeks alone on the ice. He wrote in his diary: 'All alone here now and the silence is immense'. Two weeks later, he noted: 'The Silence is so painful now that I have a continual singing in my left ear, much like a Barrel Organ, only it's the same tune all the time'.[42] Eventually, he heard the sound of a sledging song, and he thought that the solitude had finally beaten him. He rushed outside to see his four friends trudging towards him across the lonely ice sheet, and he was so excited he stood on his head. While stationed at Davis, Holliday sent a radio message to Captain Moyes, who was by then 94 years old. She told him that an ANARE geologist had just flown by helicopter to

→ Morton Moyes, a member of Mawson's AAE who endured nine weeks alone at the Western Base in 1912, later recounted his stories of 'derring-do' in Antarctica.

Gaussberg, an extinct ice-bound volcano near the Davis Sea about 377 kilometres west of Wild's Western Base. Moyes recalled that Wild's sledging parties had taken a month to get there in 1912. The geologist covered the distance in an hour.[43]

Changing times

The 1980s marked a period of modernisation on Australia's Antarctic stations when, for a time, construction workers employed by the Australian Commonwealth Department of Housing and Construction outnumbered ANARE scientists and other personnel. In the midst of this period of transition, Diana Patterson arrived at Mawson to take up her role as station leader for the 1989 ANARE. She had long nurtured an ambition to work in Antarctica and to lead a research base but, despite her experience in senior management and outdoor activities, she had applied several times without success. Finally, on the eve of the Australian Bicentennial year, the Antarctic Division selected her to serve as 'station leader elect' at Casey for the 1988–1989 summer season, in preparation for her return to Antarctica as station leader at Mawson. She would be the first woman to lead an Australian Antarctic research station.

Patterson's year at Mawson coincided with a period of profound social change that was being felt across Australian society. It was especially noticeable on Australia's remote Antarctic outposts. The masculine culture of Australian stations in those days, as Tim Bowden noted in his ANARE Jubilee history, had changed little since the first ANARE expeditions some 40 years earlier.[44] Some initially found the presence of women in Antarctica confronting, fearing that they would not be able to tolerate the physical and mental demands of living and working in such conditions. Besides, as one argument went, there were no female toilets.[45] Some suggested that women would change the culture of the bases. Such sentiments, it should be said, were not limited to Australian stations. The United States commander of Operation Deep Freeze considered the lack of suitable facilities and impact on the men acceptable reasons to exclude women from Antarctica, but he also alluded to the symbolic threat that their presence posed to the men's sense of identity: 'It was a pioneering job. I think the presence of women would wreck the illusion of the frontiersman—the illusion of being a hero.'[46] Antarctica was still perceived as a frontier and frontiers, it seemed, were no place for women.[47]

As women made their presence felt, however, tensions simmered over the appropriateness of all-male behaviour and pastimes, such as the public display of 'girlie' posters. These women, all pioneers in their own ways, were not only challenging long-held prejudices about women's capabilities on the Antarctic 'frontier'; they were also changing the cultural norms that had prevailed on Australia's remote southern outposts since the early twentieth century.[48] As Patterson later reflected:

> I had to accept the fact that as a woman I probably looked upon the world differently to my male colleagues, and I was in the minority. It made sense to me that I would have to reorientate some of my thinking and my style of management ... In the isolation of the Antarctic community, interpersonal skills were more important than assuming that authority was a given.[49]

By the end of her year at Mawson, Patterson had not only demonstrated her leadership credentials; she had also made many friends and developed an enduring passion for the Antarctic environment. The Antarctic lured her back, and she eventually served as leader at Davis station in 1995 before pursuing a career in environmental conservation.[50]

↑ ANARE biologist Maria Clippingdale assists while Lynne Irvine marks an Adélie penguin which has just been tested for infectious viruses at Davis station, 1997.

→ ANARE biologist Suzy Rea uses an Eckman sediment grabber to take surface sediment samples from lakes around Davis station, 1997.

In the field

From the earliest expeditions in Antarctica, Australians have devised various forms of shelter to protect them from the extreme conditions while working in the field. In addition to the temporary huts built at Commonwealth Bay and on the Shackleton Ice Shelf, for example, Mawson's men also cut deep trenches in the ice to serve as refuges and supply depots for the sledging parties. Aladdin's Cave was a particular favourite with the AAE men, with Mawson describing it as 'the Mecca of all sledging parties' and 'a truly magical world of glassy facets and scintillating crystals'.[51] With the beginning of ANARE in 1947, temporary field huts provided essential accommodation in some of the most remote field sites in the world.

Over the years, some field huts also became popular destinations for weekend 'jollies' away from the main stations. Casey boasts four field huts, each with its own distinctive features. Jack's Donga, about 16 kilometres from Casey, has a toilet with one of the best views in Antarctica.[52] Wilkes Hilton, originally the radio shack for the US station, has long been a favourite for expeditioners wanting to get away in their spare time. Dale Main fondly recalls his visits there: 'That was a place that you just never dreamed exist[ed]. It's a magic place. You'd get over there and it'd be minus thirty in the winter time ... and the first thing we'd do is stoke up the fire' to dry the place out before

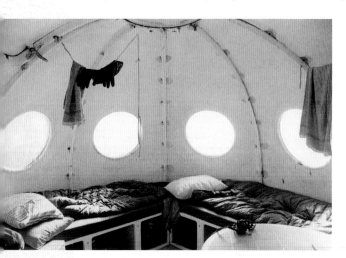

cooking and relaxing with a few beers.[53] Browning, the furthest hut from the station, is surrounded by hills and lakes and, at certain times of the year, hosts visiting elephant seals. Old Robbo's Hut at Robinson Ridge, a fibreglass hut on skis designed at the Royal Melbourne Institute of Technology, features a deck overlooking an Adélie penguin colony.[54]

← Since the 1980s, transportable fibreglass field huts have been deployed to provide shelter for Australian expeditioners undertaking research out on the ice sheet.

↑ The Apple field hut at Ace Lake.

A cluster of three prefabricated fibreglass Apple huts (red hemispherical shelters) and two Melon huts (Apples elongated with the addition of panels), constructed in 1986 in the northern Bunger Hills, became the prototype for a new type of Antarctic operation: temporary, low-cost summer camps for scientists who needed to spend extended time on the ice sheet. The Edgeworth David Base, as it became known, was located 440 kilometres west of Casey station and 85 kilometres inland from the Shackleton Ice Shelf. Expeditioners could be transported in and out by helicopter, avoiding the need for arduous traverses across the ice sheet. Other summer stations followed along the same lines and, by 2019, temporary field stations had been established at Beaver Lake (near the Prince Charles Mountains), Béchervaise Island (in Holme Bay, Mac. Robertson Land), Cape Denison (near Mawson's Huts), Law Base (Larsemann Hills) and Scullin Monolith refuge (Mac. Robertson Land) to support Australia's 'deep field activities' such as monitoring penguin and seal colonies and biological investigations of Antarctica's inland lakes.

Out in all weather

It was March 1986 when Denise Allen arrived at Mawson after being helicoptered in from the *Icebird* to work as the station's weather observer. She had already established herself as one of 'the boys from the Met. Bureau' in Queensland, and was appointed as a last-minute replacement for a male weather observer who returned home on a previous voyage.[55] Antarctica intrigued her, and she was ready for it. Growing up on farms in the postwar years, she had developed a strong sense of self-reliance and getting the job done: qualities that would stand her in good stead in the male-dominated station environment that still existed in the 1980s. Her first posting to Macquarie Island in 1985 marked the beginning of a remarkable career in Antarctica, wintering on all four permanent Australian stations and earning her the Australian Antarctic Medal. However, her first days on the continent were challenging.

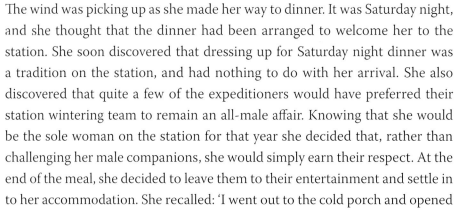

The wind was picking up as she made her way to dinner. It was Saturday night, and she thought that the dinner had been arranged to welcome her to the station. She soon discovered that dressing up for Saturday night dinner was a tradition on the station, and had nothing to do with her arrival. She also discovered that quite a few of the expeditioners would have preferred their station wintering team to remain an all-male affair. Knowing that she would be the sole woman on the station for that year she decided that, rather than challenging her male companions, she would simply earn their respect. At the end of the meal, she decided to leave them to their entertainment and settle in to her accommodation. She recalled: 'I went out to the cold porch and opened the door to find a full-on blizzard outside. It was my first blizzard, and I couldn't see my hand in front of my face'. She couldn't

↖ In 1989 Denise Allen was one of the first women to be awarded the Australian Antarctic Medal. The medal was established in 1987, replacing the British Polar Medal.

← Allen during a nine-day dog run from Mawson to Cape Bruce in 1986.

→ A 'blizz line' walk in 80 knot (148 kilometre per hour) winds at Mawson station, which averages 37 blizzard days a year.

go back to the dining hall and ask for help, because that would just confirm their fears. She recalled being told about a 'blizz line' to help her navigate between buildings during a blizzard. She was determined to find it and make her own way back to her room:

> I got outside ... and felt my way along the edge of the building. I knew I had to go uphill, then I felt the rope. I followed it and the light of my building was a welcome sight. That was my first night at Mawson, and I never told anyone about it until September![56]

On the following morning, she learned that the *Icebird* had been unable to complete its resupply mission and was returning to Hobart, along with most of her clothing. She wore some men's clothing for the rest of the year. By the end of her first season in Antarctica, Allen was hooked: 'I loved the place and I still love the place'.[57]

Widely considered the most inhospitable of Australia's three continental stations, Mawson's notorious winds posed particular challenges for weather observers like Allen, who had no choice but to venture outside to undertake their shifts. She recalls one occasion when the wind was howling at 70 knots (130 kilometres per hour):

> I found out that you could walk unaided up to about your weight limit, and I was 54 kilograms, so I knew I was going to be struggling. I thought I'd have a go at releasing my balloon, so I went into the balloon shed and opened the doors. With the first balloon off the tray—the suction pulled me along and I knew that if I kept on hanging on it was going to be stretched and ripped by the wind. So I let go and it blew out into the snow. I shut the doors and went out along the blizz line to the radar dome to check the radar screen and see if it was a successful release—and it was! I was almost jumping for joy. The wind was gusting to about 95 knots [176 kilometres per hour] at that time—that was my most successful balloon release.[58]

Allen's shift usually began at 4am, and she can recall the spectacular displays of the aurora australis in the night sky:

> *I saw my first aurora at Macquarie Island. I was coming back from a night shift, and it was dark. It was like a lace curtain blowing in the winds … It felt like I could just touch it. It was different at Davis, as though a paint pot had been flung out over the sky—all purples and pinks and greens.*[59]

She was also fascinated with the appearance of nacreous cloud, an 'oily', iridescent mother-of-pearl cloud produced by ice crystals in the stratosphere. It was a sign of extremely cold temperatures in the upper atmosphere that could be as low as −80°C.

↓ ANARE members enjoying the last husky run at Mawson station in 1993 before the ban on introduced species came into effect under the Protocol on Environmental Protection to the Antarctic Treaty (the Madrid Protocol).

In addition to her work as a weather observer, Allen treasured her times away from the station, running with the dogs. She recalls one particularly memorable dog run from Mawson to Cape Bruce:

The dogs were so good for morale. Everybody took turns with them and if you got off station, even for an hour or two, people would come back chatting about the dogs … The sound of being out there in the wilderness—it is absolutely silent apart from the sound of the sled on the snow, and the dogs panting with their tails up in the air as they're running, and you are on the sled behind them or running beside them with the sled moving across the ice—it is a really pleasurable feeling … you are just self-sufficient, surviving.[60]

On the way back to Mawson, the sledging party came across a new tide crack near a glacier. As they were trying to find a way across, the dogs caught the sound of a seal under the ice beneath a newly frozen tide crack. They went berserk. The just-formed sea ice was still dangerously thin so, after gaining control of the dogs, the party decided it was safer to head back to the station than risk another night on the ice. During their nine days away, the group of four managed to cover 200 kilometres, sometimes running with the dog teams and sometimes pushing the sledges. As Allen recalls, she came back from that dog run lean and fit.

Allen eventually wintered at all four of Australia's permanent stations during the 1980s and early 1990s. Returning to Mawson and Davis a decade or so later, she observed how much had changed. There were vast improvements in the food and telecommunications technology, but the upgraded station facilities also created more opportunities for private activities so that the expeditioners were less inclined to socialise with each other. The administration of the station had changed too. Decisions that used to be made by the station leader were increasingly referred to Head Office in Hobart, while laws regarding matters such as health and safety, discrimination and sexual harassment were strictly enforced. While most of the changes were clearly for the better, Allen admits that she sometimes missed the adventurous spirit of the earlier years.

← Nacreous clouds are more commonly seen in Antarctica, but can also be observed in the Northern Hemisphere.

Unless one has actually seen it, it is impossible to conceive the stupendous extent of this ice-cap, its consistency, utter barrenness, and stillness, which sends an indefinable sense of dread to the heart. There is nothing beautiful to contemplate, no contrasts, absolutely no diversity, but for all that it is majestic and affords a profitable theme for meditation.

LOUIS BERNACCHI[1]

Inland

Exploring the ice sheet

← The Transantarctic
Mountains, 2021.

The earliest expeditioners traversing the Antarctic ice sheet had relied entirely on dog teams or manpower to haul heavily loaded sledges with essential scientific equipment and precious supplies across the ice. In the postwar years, however, Antarctic expeditioners became the beneficiaries of new technologies that had been developed to support military operations in snow conditions.[2] With the establishment of Mawson, Wilkes and Davis research stations, Australian scientists began to make extended forays into the interior to take seismic soundings of the ice sheet and examine its physical structure.[3] Even after 60 years of scientific expeditions, little was known about the nature of the Antarctic ice sheet. Glaciological research was only just beginning in earnest, but Phillip Law could see its potential to answer crucial questions about whether the continental ice sheet was expanding or shrinking. He later wrote:

> *Here the processes that scarified the surface of the Earth during the great Pleistocene Ice Age are still proceeding. For the first time, man has had an opportunity to study a vast continental ice sheet in situ.*[4]

During these early forays onto the ice sheet, scientists were intent on gathering data about its altitude, depth, composition, structure and stability. They especially wanted to understand how and where snow accumulated, and how much was blown or melted away.

← Australian geologist Arch Hoadley, measuring snowfall during Mawson's AAE in 1912.

→ The practice of manhauling sledges across the vast Antarctic interior became a defining image of the 'heroic era' of Antarctic exploration.

Mawson's huskies

For much of the twentieth century, working dogs played a key role in enabling humans to survive in Antarctica. The British Antarctic Expedition was the first to bring dog teams to the continent in 1898. The huskies, originally from Greenland or Labrador, were ideally suited to Antarctic conditions, having the strength and stamina necessary to haul heavy sledges across the ice. Douglas Mawson recognised the importance of dog teams for his AAE and, in 1911, secured 48 Greenland dogs for the voyage aboard the *Aurora*. Many succumbed to the terrible conditions at sea and Mawson was forced to bring another 38 dogs on board in Hobart. Once in Antarctica, the dogs suffered in the biting winds and snowdrifts, but proved their worth in the extreme conditions. Mawson wrote:

> There can be no question as to the supreme value of dogs as a means of traction in the polar regions. It is only in such special circumstances as when travelling continuously over very rugged country, over heavily crevassed areas, or during unusually bad weather that man-hauling is to be preferred … where human life is always at stake, it is only fair to put forward the consideration that the dogs represent a reserve of food in case of extreme emergency.[5]

Dog teams were introduced to Mawson station in 1954 and became a much-loved feature of life there for nearly 40 years. Expeditioners would proudly record their tally of 'dog miles' covered during sledging trips. In one year alone, the Mawson team covered 6,084 dog miles (9,791 kilometres). The longest dog-sledge journey in ANARE's history happened in 1958 when surveyor Graham Knuckey, geologist Ian McLeod and radio operator Peter King flew to Amundsen Bay and sledged back to Mawson, a distance of 650 kilometres. Glaciologist William (Bill) Budd fondly remembers the husky teams and recalls how they withstood the extreme conditions during a particularly difficult traverse to the Vanderford Glacier, where they were able to negotiate a large area of crevasses that would have been treacherous for heavy vehicles.[6] Dogs were also a source of fun around the station: 'morale boosters *par excellence*!', according to Rod Mackenzie.[7]

It was a sad day in 1993 when the last six huskies were removed from the continent, following a ban on introduced species under the Protocol on Environmental Protection to the Antarctic Treaty (the Madrid Protocol). Nineteen of the younger dogs and three pups born at Mawson began a new life as working dogs in the United States. Two dogs, Morrie and Ursa, were inseparable in Antarctica and are still together in death in Museum Victoria's collection.

Dogs are no longer part of life on Australia's Antarctic stations, but their spirit lives on. In 2004, when the Australian government purchased two CASA212 aircraft to transfer people and equipment between stations, school children named them 'Gadget' and 'Ginger' after two of Mawson's beloved huskies. In 2017, recognising their loyal service to the AAE, the Australian Antarctic Division Place Names Committee gave dog names to 26 islands, rocks and reefs off Cape Denison.[8]

Another dog in Antarctica has also won hearts. In the 1990s, Stay, a fibreglass donation dog for Guide Dogs for the Blind, found her way onto Australia's Antarctic stations after being smuggled aboard the *Aurora Australis*. Stay arrived just as the working huskies were being removed. She remains a cherished, if unofficial, member of Australia's Antarctic program and is regularly photographed at locations around the continent and subantarctic islands. She has wintered at every Australian station, visited bases and ships, and made her way to the North Pole where she posed for a photograph at Roald Amundsen's memorial. The adventures of Stay have become legendary, and regularly feature in news updates on the Australian Antarctic Division's website.[9]

↗ Husky teams based at Mawson station between 1954 and 1993 not only proved their worth on sledging trips, they also provided much enjoyment and companionship for those living and working in this remote region.

← 'Stay' on duty at Proclamation Point, Mawson station. The stories of Stay's quirky adventures have captured many hearts in the Antarctic community.

↑ In the centre of the Antarctic ice sheet lies Russia's Vostok research station.
In 1983, it experienced the lowest temperature ever recorded on Earth (−89.2°C).

The Vostok Traverse

One of the most celebrated Australian traverses of the ice sheet took place in 1962–1963 when New Zealander Robert (Bob) Thomson led an expedition team comprising an American meteorologist, Danny Foster, and four Australians (glaciologist and photographer Alastair Battye, geophysicist Don Walker, driver and mechanic Desmond 'Pancho' Evans and senior mechanic Neville 'Gringo' Collins). The expeditioners set out from Wilkes bound for the remote Vostok station, high on a plateau in the heart of Antarctica, thought to be the coldest place on Earth.

The Vostok party was well equipped for the traverse. The team had two of the three station Weasels and two Caterpillar D4 tractors towing nine sledges, including two mounted with caravans. Nevertheless, the men made slow progress across the polar plateau. Undertaking fieldwork in such isolated and extreme weather conditions was a daunting and risky venture, even in summer. Frequent blizzards and white-outs restricted travel to one day in every three and, after each blizzard, the men were obliged to dig the vehicles out before they could continue their journey. Sastrugi also regularly brought proceedings to a halt and, to make matters worse, the air temperature kept falling. The party measured the lowest temperature of the traverse at –63.4°C on 27 October. Neville Collins can remember the extreme cold, and how quickly cheeks and noses would become white with frostbite[10]:

> *Minus sixty was a sort of average day. After minus forty it doesn't seem to feel much worse. Up to minus forty you can feel it's cold, but after minus forty you don't feel a great difference. You notice when you breathe out the air crackles and makes this annoying crackling noise below minus forty.*[11]

Apart from the intense cold, the men experienced the strange sensation of being enveloped in the vast white ice sheet for weeks on end. As Thomson recalled, 'it was just like having one's eyes six inches [15 centimetres] from a large white sheet'. Out on the interminable ice sheet, time and space condensed into a meaningless abstraction.[12]

While the little Weasels braved the vagaries of the ice sheet and Antarctic weather, they demanded frequent maintenance and large supplies of fuel. For the first 480 kilometres, the traverse party relied on fuel drums dumped at one mile (1.6 kilometre) intervals during the previous summer, supplemented by a fuel drop at latitude 74° S by a United States Air Force plane. Apart from keeping themselves and the Weasels moving, the team also had their scientific tasks, and mapping the characteristics of the ice sheet and bedrock was one of the most important. It was a relatively unsophisticated affair. Using a tractor to drive the drill, they would bore deep into the ice, lower thermometers and take the temperature of the ice sheet at different levels, each measurement offering a glimpse into Earth's past climatic conditions. Then, with a series of microphones laid around the hole at the surface, they would lower dynamite into the hole and detonate it, recording the soundwaves as they travelled down through the ice from the site of the explosion, struck the bedrock and bounced back. After the blast, they would analyse the data transmitted to equipment on the surface and use it to calculate the depth of the ice.[13]

↑ The traverse train at Vostok station in 1962.

After two months of traversing nearly 1,500 kilometres, the party rumbled into the abandoned Soviet station at Vostok. The Soviet authorities had recalled its occupants a year earlier and, by the time the Australians arrived, the buildings were buried in snow. The group managed to climb in through a hatch in the roof. In the living quarters, they found a table laid out for three, a pan of frozen steak and onions on the stove, and a pot of tea. The Russians had apparently left in a hurry. The new arrivals simply turned on the generator, reheated the food and sat down to one of the best meals of their four-month journey.[14]

Deep field research

The Vostok Traverse heralded a new era of long-distance traverses on the ice sheet.[15] The journey was a remarkable one but, more importantly, it demonstrated the feasibility of undertaking long-distance tractor-train travel across the ice sheet and showed that it was possible for Australian scientists to conduct fieldwork away from the stations for extended periods. Traverses also offered expeditioners a temporary escape from the confines of the station, and a chance to experience 'life in the raw'. As Stephen Murray-Smith observed, traverses may not have involved sophisticated science, but Australia was very good at them.[16]

Keith Gooley realised his ambition to visit Antarctica in 1971 when, armed with an electronic engineering degree and experience in the Australian government's Ionospheric Prediction Service, he found himself on the way to Mawson station to work as an ionospheric physicist.[17] His job was to install and operate equipment for measuring the ionosphere and how it affected radio wave propagation, and he has vivid memories of working in the many blizzards that year. Gooley was required to make his way between the various buildings at Mawson, but going outside in such conditions was always a high-risk activity. First he had to remove his glasses, before making his way blindly along the 'blizz line' that ran between the buildings through the swirling snowstorm. He knew the dangers, having heard stories from others who had been blown downwind when they accidentally let go of the line and were forced to work their way back through the full force of the wind. There was also the ever-present risk of fire. Gooley recalls the system of night-watch in those days, when one expeditioner would be rostered to stay up all night and tour the station every three hours to keep watch for any sign of fire:

> *Fire was, and probably still is, a very serious hazard, a lot more so than here because it is very dry and cold, and the relative humidity is extremely low. When you warm that air up it has little moisture in it.*[18]

One of the highlights of Gooley's year at Mawson was going out onto the ice sheet by tractor-train to drop fuel at a depot for a traverse to be undertaken to the Prince Charles Mountains in the spring. He recalls climbing up Mount Twintop to get a better view of the plateau:

→ Keith Gooley in the living caravan 'Brigadoon' near Cape Folger about 40 kilometres east of Casey station in 1974.

You look out over this expanse of unending snow and ice with just the occasional mountain ranges sticking up through the ice. It's magnificent, just this white shimmering wilderness. It's beautiful, and you can look down into the crevasses—you get all the shades of blue right through to black. The experience lives with you. On a nice day, when there's no wind and the sun's shining, it's absolutely glorious. With no trees to rustle in the breeze and no birds twittering, there is dead silence. You can just hear the blood rushing in your ears.[19]

→ For all its reputation as a barren wilderness, the experience of being in Antarctica leaves few unmoved.

↑ Mackenzie Bay at the western extremity of the Amery Ice Shelf,
named for the master of the BANZARE ship *Discovery* in 1930–1931.

Over the next few years, Australian researchers spent the summer months studying
the Amery Ice Shelf, the largest ice shelf in East Antarctica and one of the region's most
spectacular features. Studies by Soviet glaciologists in the 1950s had shown that the
heart of the Antarctic ice sheet was moving very slowly, but that the ice at the edges
was moving more rapidly.[20] In 1968, an Australian team spent the winter camped on the
Amery Ice Shelf, aiming to measure the speed and flow of ice from the Lambert Glacier
using a series of marker poles installed on previous field trips. Lambert Glacier—the
world's largest glacier—is estimated to be about 40 kilometres wide and 400 kilometres
long and drains around ten per cent of the Antarctic ice sheet, calving massive icebergs
into Prydz Bay. The movement of the glacier was responsible for a depression in the
East Antarctic Ice Sheet, exposing an area of bedrock known as the Prince Charles
Mountains.[21] The 1968 team, comprising Max Corry (leader), Neville Collins
(a veteran of the Vostok Traverse), Alan Nickols and Julian Sansom, established a camp

100 kilometres from the front of the Amery Ice Shelf to avoid floating out to sea if the shelf broke away from the continent. The group soon found that they were exposed to ferocious katabatic winds that blew almost constantly down the shelf. Heavy snowfalls added to their plight, burying their caravans, and forcing them to dig a network of underground tunnels to provide access to the various parts of the camp. The 'Amery Troglodytes', as they became known, worked throughout winter using a thermal drill to penetrate the ice shelf, producing ice cores that gave the first glimpses of the physical characteristics of the Antarctic ice sheet. They found that the sheet as a whole was moving coastward at about ten metres per year; the Amery Ice Shelf was travelling more rapidly. While Mawson's AAE had sought to understand the origins of the continent by studying the rocky outcrops protruding from the ice, it was becoming clear that the ice held secrets of its own.

The 'jewel in the crown'

In the summer of 1984–1985, Australian palaeontologist Patrick Quilty made an astounding discovery. He was on a field trip aiming to obtain a geographical fix on the Vestfold Hills near Davis station in order to adjust the charts that had been prepared from aerial photographs taken by the United States during Operation Highjump in 1947. During the reconnaissance, Patrick and his companion stumbled on fragments of fossil bone. As he recalled:

We were wandering around the edges of one of the lakes, and then about 10–15 metres away I saw a pile of fragments on the ground ... I went over ... and put a bit of it under the hand lens ... and it was bone. I immediately recognised the significance of it, because here was vertebrate material, 3.5–4 million years old. The only other vertebrate material known from Antarctica is 40 million years old or more ... I didn't know what I'd found ... A couple of larger ones that told me it wasn't just a piece of leg bone or a rib, it was something interesting. And so I wrapped these ... sketched the area, photographed it, numbered all the large fragments, re-photographed it, collected it carefully in bags. I brought this lot back to Australia—and it turned out to be the first of the dolphins.[22]

In the area known as the Marine Plain, he had found the remains of dolphins belonging to a species and genus previously unknown to science.[23] The plain was an ancient seabed littered with the skeletal remains of single-celled algae, known as diatoms, that had drifted down as the sea ice melted and settled in thick yellow sediments on the seabed. When Stephen Murray-Smith visited the area during his trip to the Australian stations the following summer, he thought it an ugly landscape. 'It is the ultimate abomination of desolation,' he wrote in an article for *The Australian* newspaper, 'yet it holds important clues to life on earth.'[24] The skeletons discovered by Quilty were around four million years old and remarkably complete. After death, it seemed that they had drifted down to the seabed and remained, undisturbed, in the Pliocene-era sediments. It was the only record of a vertebrate animal in the last 40 million years of Antarctic evolution, and the only place on the continent where the rocks remained undisturbed by geological processes. Quilty's research earned him international acclaim, and he became a leading figure in international and Australian Antarctic science, with awards including Member of the Order of Australia (AM) in 1997, the Phillip Law Medal in 2016 and the Australian Antarctic Medal, awarded posthumously in 2020. Quilty Bay in the Larsemann Hills (near Davis station) and Quilty Nunataks (West Antarctica) were named in his honour. Quilty's discovery provided crucial evidence that Antarctica had been much warmer than the present, and that the ice cap was younger than previously thought.[25] Murray-Smith declared Marine Plain to be the 'jewel in the crown' of Australian Antarctica.[26]

← Fossils left behind from a past geological expedition at the protected site of Marine Plain near Davis station.

→ Marine Plain, with its rich fossil record, has been described as the 'jewel in the crown' of the Australian Antarctic Territory.

Traversing the ice sheet

Until the development of motorised snow vehicles after the Second World War, expeditioners relied largely on wooden sledges—similar to those designed by indigenous peoples for use in the Arctic—to haul themselves, their equipment and supplies across the ice sheet.[27] While some early expeditioners hauled the heavy wooden sledges using ponies or manpower, Mawson decided to use dogs bred and trained in Greenland for the Australasian Antarctic Expedition (AAE). He also planned to make use of newly available aviation technology to explore Antarctica's vast geographical region, although his specially imported two-seater Vickers monoplane crashed during a test flight in Adelaide. Not to be defeated, Mawson decided to use the fuselage and engine to fashion a motorised sledge, with the intention of using it to tow sledges while laying depots for the AAE's summer sledging parties. Once at Commonwealth Bay, Frank Bickerton undertook the conversion from monoplane to sledge. The resulting air-tractor, fondly known as the Grasshopper, made a successful trial run to a depot known as Aladdin's Cave, but the engine failed while towing four sledges and the three members of the Western Party in 1912, and the men were forced to abandon it and haul the loaded sledges themselves. The air-tractor was subsequently brought back to the Main Base where it remained, buried in the ice, until Mawson's Huts Foundation located some of the remains in 2008.

By the 1960s, sledging had largely given way to motorised caravans for long-distance traverses, hauled by Weasel tractors fitted with tracks. A product of military engineering, Weasels were used extensively by the US Army and deployed in the Arctic, before being sold off in the 1950s for civilian use. Designed for extreme conditions, they were capable of hauling heavy loads in soft snow over long distances, in arguably greater comfort and safety than sledges, and they had the added advantage of being amphibious.[28] This new form of oversnow vehicle provided shelter, heating and insulation,

→ Sydney's Powerhouse Museum cares for a collection of wooden sledges and other gear used by members of Douglas Mawson's AAE in 1911–1914.

→ Australia's Antarctic aviation system transports passengers and cargo between Hobart and Wilkins Aerodrome near Casey, while smaller fixed-wing aircraft and helicopters link the continental stations and support scientific fieldwork.

enabling a field party to undertake scientific and survey work without having to expend energy in keeping themselves alive, even in a blizzard. Once inside the insulated cabin, even protective clothing was no longer necessary.

Methods of oversnow transport continued to evolve to tackle the extreme physical conditions experienced on the ice sheet. The Australian government eventually replaced the Weasel with modern tracked vehicles for long-distance fieldwork, including the Swedish dual-cab Hägglunds and the Canadian Nodwell and Foremost Pioneer. Air transport also became a crucial part of Australia's Antarctic activities at this time, with helicopters being used for ship-based reconnaissance flights over sea ice and deploying people and equipment from ship to shore. They were also invaluable for short-distance journeys across the ice, ferrying expeditioners between stations or to fieldwork sites. In 2003, the Australian Antarctic Division introduced fixed-wing aircraft to extend the range of its field activities, transferring 22 expeditioners and scientific cargo from Davis station to relieve the wintering party based at Mawson station. The journey took five hours to cover the 650 kilometres between stations, a journey that would have taken several days by ship.[29]

During his year as leader of the ANARE program at Mawson in 1959–1960, John Béchervaise reflected on how much transportation practices had improved since his first experience of Antarctica just five years earlier. The plateau was 'about as incapable of supporting human life ... as the surface of the moon', but expeditioners could now travel and live in specially equipped cabins mounted onto cargo sledges. Antarctic exploration had become an exercise in efficiency rather than discovery. Perhaps, he pondered, even the elderly and frail might one day be visiting the South Pole.[30]

*[The Southern Ocean] touches the lives of us all ... a living,
breathing, organic body of water that embraces us and nurtures
us in ways we cannot yet fully appreciate.*

TONY PRESS[1]

Ocean

← An iceberg adrift in the Southern Ocean.

The circumpolar Southern Ocean with its seasonal freeze and thaw is as much a part of the Australian Antarctic story as the continent itself. In 1776, when James Cook anchored in Bird Bay (Christmas Harbour) in the Kerguelen archipelago during his second voyage of exploration in the Southern Hemisphere, he marvelled at the plentiful penguins and fur seals that were so tame that his men were able to take as many as they pleased. In the wake of his voyage, sealers and whalers wasted no time in making their way south to these promising new fishing grounds in the Southern Ocean. It would be nearly a century, however, before scientific expeditions began to explore this little-known region.

In 1872, the *Challenger*, a three-masted, square-rigged warship, set sail from the English town of Sheerness. It was to be the first voyage of its kind, a scientific journey around the globe to investigate the physical and biological characteristics of the world's oceans. Under the command of George Strong Nares, the *Challenger* carried 20 officers, 200 crew members and a team of six civilian scientists led by a professor of natural history at the University of Edinburgh, Charles Wyville Thomson.[2] The *Challenger* spent four years tracking across vast expanses of ocean, building a massive collection of samples and recording data from the depths of all the major ocean basins. The vessel reached as far south as latitude 66° S and, on 16 February 1875, those on board toasted the crossing of the Antarctic Circle. Nevertheless, the Southern Ocean was always going to be the most challenging of oceans for scientific research. It would be another 40 years, during the pioneering polar expeditions in Antarctica, that the Southern Ocean became the subject of closer investigation. When Mawson's Australasian Antarctic Expedition (AAE) set sail from Hobart in 1911, the hearts and minds of the expeditioners were focused on unravelling the mysteries of the new continent to their south. While it received much less public attention, another party of scientists would be undertaking marine research from aboard the *Aurora* under the command of Captain John King Davis.[3] As Davis noted:

← The *Challenger* off Cape Challenger, Kerguelen Islands, in 1874 during the four-year scientific expedition to investigate the depths of the world's major oceans.

↑ Scientists aboard the *Discovery* inspecting a haul dredged from the Southern Ocean during BANZARE in 1929–1931 (from left: Douglas Mawson, William Ingram and James Marr).

Very little was known of the seafloor in this area, there being but a few odd soundings only, beyond a moderate distance from the Australian coast. Even the great Challenger expedition had scarcely touched upon it; and so our Expedition had a splendid field for investigation.[4]

Standard oceanographic textbooks of the day were silent on how to deal with the practical difficulties of operating sounding and trawling equipment in the turbulent waters of the Southern Ocean, so Davis devised his own methods based on advice from Dr W.S. Bruce, leader of the Scottish National Antarctic Expedition, and his own observations of trawling operations in Australian waters. To obtain a deep-sea sounding in heavy swells, for example, Davis would steady the vessel and use the engines to counter any drift while adjusting the sounding wire to prevent it from snapping whenever the ship rolled violently. The expeditioners also measured sea temperature and salinity, calculated the depth of the continental shelf surrounding Antarctica, mapped the contours of the seafloor south of Australia and collected specimens from the deep.[5]

← Scientists measuring the albedo of Antarctica's sea ice to determine its ability to reflect the sun's energy, circa 1988.

From these modest beginnings, the ocean sciences would emerge as an integral part of Australia's Antarctic program. By the late twentieth century, efforts intensified to understand the nature of the Southern Ocean's distinctive ecosystem and its role in shaping global ocean circulation and climate patterns.[6] At the same time, ice cores retrieved from the Antarctic ice cap during the 1990s were yielding compelling new evidence about the rhythm of past climatic fluctuations. The impetus for intensive scientific investigation of the Southern Ocean came in part from a revolutionary theory developed by American geoscientist Wallace (Wally) Broecker in the 1980s. Citing evidence of climate history retrieved from cores drilled in the Greenland ice cap, he proposed that there was a direct link between ocean circulation and climate change. His theory explained how Earth's systems were interconnected in a delicate dance of ice, temperature, ocean currents and vegetation, and he coined the term 'great ocean conveyor' to explain how sudden changes in the circulation of deep-ocean currents had triggered past ice ages.[7]

In 2006, the Integrated Marine Observing System commenced in Australian waters and, in the following year, the International Council for Science, the World Meteorological Organization, the Arctic Council, parties to the Antarctic Treaty and other international organisations combined forces to examine the influence of the two polar regions on climate as part of the fourth International Polar Year (IPY).[8] When Australian and American scientists collaborated in a joint research program to examine the powerful westerly winds of the high southern latitudes in 2010, they found that the winds were strengthening and creating changes in the surface layer of the water, with implications for the way in which the ocean and the atmosphere exchange heat and carbon dioxide.[9]

↑ Penguins keep their balance on an iceberg off the South
Shetland Islands as it is swept by huge Southern Ocean swells.

← Autonomous underwater vehicles show how far ocean research has come since Mawson's day, enabling scientists to gather information about the mysterious world beneath Antarctica's sea ice.

Two years later, an international team of oceanographers joined forces to establish the Southern Ocean Observing System, which examined the role of the ocean's deep currents and water masses in shaping global climate patterns and cycling carbon and nutrients. By deploying autonomous underwater vehicles and conductivity–temperature–depth (CTD) profilers to continue the time-honoured tasks of measuring depth, temperature and salinity, ocean scientists were able to take the pulse of the ocean, from the deepest Antarctic bottom water to the surface. They also began to track the circulation of ocean water around the globe, mapping changes in the physical, biological and biogeochemical properties of the ocean beneath sea ice, revealing how deep water from the seafloor interacts with the atmosphere to regulate global atmospheric temperatures. In the process, Australian oceanographers discovered that the normally dense bottom water of the Southern Ocean was considerably less dense than it had been in the 1970s, and that this was reducing the capacity of the ocean to store carbon dioxide. The Southern Ocean, like the Antarctic continent itself, has become a sentinel of climate change.[10]

Stories from the seabed

When deep-sea coral biologist Dr Narissa Bax embarked on her first voyage south aboard the Australian resupply ship *Aurora Australis* in 2009–2010, she was contributing to an Australian Antarctic project aimed at understanding the impact of fishing on the seafloor environment in the Southern Ocean.[11] The first destination was Bruce Rise, a previously unexplored part of the seabed off the East Antarctic coast, about 230 nautical miles northwest of Australia's Casey station. Here, the marine scientists lowered trawl nets to collect samples of marine creatures, and cameras for capturing images of the nature of the benthic environment on the seafloor. Apart from the technology and equipment being used, the process of sampling and measuring was little different to that used by Mawson and his scientists aboard the *Aurora* a century earlier.[12] As the cameras probed the seafloor, Bax and her colleagues were surprised to find a muddy plateau, superficially devoid of life apart from sea cucumbers and worms, and the distinctive markings of what appeared to be longline fishing trawls imprinted in the sediments. It was a stark reminder of the impact that humans have had, even in the depths of these remote and little-known waters.

After three days, the *Aurora Australis* relocated to Tressler Bank off the Shackleton Ice Shelf between Casey and Davis stations. Here, the trawl-mounted cameras revealed a diverse terrain, including a steep shelf and rocky outcrops, supporting a rich array of invertebrate species such as sponges and corals. In areas of the seabed less impacted by ice scour, Bax found the longer-lived species of corals. Previous research showed that this part of the seabed, near the Mertz Glacier, supported beautiful fields of hydrocorals,

↓ The *Aurora Australis* heads south for Casey station during the second voyage of the Antarctic summer season in 1997.

AQUARIUM PLANT

↑ The aquarium on the *Nuyina* was specifically designed to house Antarctic krill, and other more fragile creatures, captured in *Nuyina*'s unique 'wet well'.

← Rob King designed the marine research facility aboard the *Nuyina* to improve on earlier methods of trawling and increase the chances of capturing and studying live krill with minimal disturbance to the fragile creatures.

or lace-corals, recognised as an important habitat for other species. This part of the East Antarctic seabed was identified as vulnerable during the Census of Antarctic Marine Life, a ten-year international project to assess the diversity, distribution and abundance of marine life in the world's oceans.[13] As samples of the variety of marine life were brought to the surface, Wendy Pyper, a science journalist from the Australian Antarctic Division, described the scene in the 'wet lab' on board the vessel:

> I help sort some of the first tubs of 'creepy-crawlies' brought into the wet lab, picking out different species with forceps and putting them into individual containers filled with sea water for the biologists to classify, weigh and photograph. The work is back-breaking as we jostle for space, hunched over long sinks, gently untangling long-armed brittle stars and feathery hydrozoans and sifting through shards of coral, shells and grit for hidden gems. While some invertebrates don't take kindly to being crushed by hundreds of kilograms of their compatriots, many are surprisingly resilient. Amphipods, for example, like the 'beach fleas' or 'sandhoppers' on the beach, have a tough and often ornate exoskeleton which, along with their small size, allows them to weather the worst of conditions … After five days of biological nirvana, the biologists have catalogued and photographed about 430 species.[14]

The vessel's first resupply stop was at Casey, giving Bax a rare opportunity to go ashore for the day during the 'changeover'. She volunteered for 'slushy' duties, then spent the rest of her time absorbing the atmosphere of the isolated station:[15]

→ The ship's innovative aquarium allows Australian scientists to study the impacts of fishing and climate change on krill in the Southern Ocean.

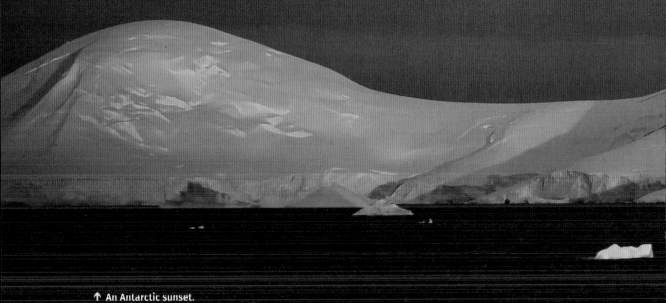

I love it so much. The experience of the aurora australis when you're at sea for five or six days, and then you're in the ice and you have those amazing sunsets and sunrises. I used to stay up to see the sun set over the horizon and watch it come back up. Then as you go through the ice you see the krill coming out of the ice. You can be surrounded by 50-odd orcas, the seals, the seabirds, the icebergs. The feelings of awe and happiness come to life in Antarctica. You're very insignificant amongst it all, and you're very present and don't want to miss anything. It's all about viewing your natural environment, and you become hyper-aware of things. You notice that when you've been away for a while and come back to Tasmania, you can smell the eucalyptus.[16]

↑ An Antarctic sunset.

In her line of work, however, Antarctic expeditions were all about the sea, studying the benthic environments along the continental shelf and marvelling at the great beauty and diversity of Antarctic corals. During one night trawl, she recalls how the creatures brought to the surface were glowing with bioluminescence:

> There's a really emotive, beautiful animal—the crinoid—that looks as though it is dancing. When they come up in the trawl they are still alive and their arms keep moving ... There's a lot of emphasis on keeping these animals intact and alive, and to photograph them in their natural form before they are preserved and lose their colour and important characters for identification.[17]

When Bax sailed south again in 2016 with the Antarctic Circumnavigation Expedition vessel *Akademik Tryoshnikov*, she was part of an international project studying benthic marine life along the continental shelves around Antarctica. The team members already knew that the Southern Ocean, with its vast expanses of open water, absorbed more heat and carbon dioxide than any other part of the world, but they wanted to understand the role of the Southern Ocean in cycling carbon, and how much carbon it takes with it to its final resting place on the seabed.[18] In those chilly waters they found microscopic diatoms and other phytoplankton that inhabit the briny veins of sea ice and stain it yellow, serving as food for Antarctic krill (*Euphausia superba*). These organisms also carry out almost half the photosynthesis on Earth, absorbing carbon dioxide into their delicate skeletons.

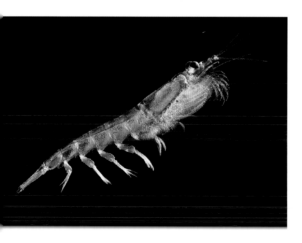

When they die, the skeletons float down through the water column, taking the carbon dioxide with them to the ocean floor where they join the seabed animals, such as corals, clams and bryozoans. The remains of these marine creatures eventually sink into the mud, locking up the carbon dioxide in the sediments of the continental shelf of Antarctica for thousands of years.[19]

← Antarctic krill are at the heart of the Southern Ocean food chain and play a crucial role in removing carbon from Earth's atmosphere

Bax and her colleagues were interested in the impact of increasing acidification—a chemical change caused by increasing levels of atmospheric carbon dioxide dissolving in the ocean—on marine species with shells made from calcium carbonate. Acidification will have catastrophic effects on these species, reducing the amount of carbonate available to make their shells, and lessening their ability to trap carbon dioxide in the depths and slow the rate of climate change. The seabed around Antarctica continues to yield new species, and there is increasing pressure on marine scientists to identify them and to better understand how they are reacting to the impacts of climate change in the Southern Ocean.

Life on the edge

Polar sea ice is one of the largest ecosystems on Earth. All living things that inhabit the cold polar waters around Antarctica rely on the seasonal ebb and flow of sea ice. Ice is the language of this region—not only the vast ice sheet and glaciers of the continent, but also the ocean that surrounds it. As I wrote in *Wild Sea*:

> *Sea ice is neither land nor sea but something else entirely. Sea ice is the littoral, the shoreline, of Antarctica. It dictates everything here—movement, temperature, colour, life and death—screeching and grinding and screaming in protest as the ship's strengthened hull forces through it a narrow path.*[20]

Sea ice forms as the ocean surface freezes over, taking on different shapes and colours as it thickens and expands into ever larger pieces. Apart from the tabular icebergs that calve from the ice shelves and glaciers, sea ice is the most distinctive feature of the Southern Ocean. In winter, it covers 17 to 20 million square kilometres—more than half of the ocean's surface. In summer, it melts away to around three to four million square kilometres, breaking down into icefloes of different sizes and shapes. Stories of navigating the sea ice surrounding the continent are as legendary as the sledging journeys to explore the ice sheet within it. Louis Bernacchi was spellbound by it as the *Southern Cross* neared the Antarctic coastline in 1898:

*In all Nature's realm
there are few sights
more impressive
than a vast field
of magnificent
glittering ice-floes
on a beautifully calm
morning with the deep
blue Antarctic sky
overhead. Lonesome,
and unspeakably
desolate it is, but
with a character
and a fascination
all its own.*[21]

← Frank Hurley captured
this image during the
BANZARE expedition in
1929–1931, describing it
as: 'A crystal garden in
Antarctica. An enchanting
glimpse of shattered
icefloes and rosette ice
crystals in the heart of
the icepack'.

Sea ice, as he soon discovered, is also unpredictable. After a few days of making little headway, the ice closed in around the ship and the vessel was imprisoned for 43 days. Few visitors had experienced the character of the sea ice at that time, and Bernacchi speculated about its behaviour. He noted that it froze to an average depth of four to five feet (1.2–1.5 metres) and extended out into the ocean for around 50 miles (80 kilometres), until the 'perpetual violent agitation' of the Southern Ocean's currents and winds prevented its formation.[22]

Observing and recording the movements and behaviour of life above and below the sea ice became a significant part of Australia's scientific work in the region. The biological riches to be found along this forbidding coastline have provided expeditioners on the stations with the opportunity to study Antarctic plant and animal life at close quarters.[23] Pack ice surrounding the Antarctic coastline appears solid, but it is constantly moving with the winds and currents of the Southern Ocean, as Shackleton discovered when the *Endurance* was caught and eventually crushed. Fast ice, where sea ice is anchored to the coastline, is thicker and forms ice shelves and grounded ice.[24] When Tracey Rogers visited Antarctica as part of her doctoral research to study leopard seals (*Hydrurga leptonyx*) that inhabit the edge of the fast ice, she thought the ice edge 'the most beautiful place'.

> *It's where it's all happening—and you've got the sounds of it all, the creaking and the seals singing. I always record at night so it's all pink, and the birds—the penguins—think that you're a flock of penguins and they're all around you and giant petrels swoop on something dead on the ice, and the leopard seals come to check you out to see if there's something to eat. It's a really lovely, lovely place, it's really special.*[25]

→ Ernest Shackleton's objective during the *Endurance* expedition was to cross Antarctica from the Weddell Sea to McMurdo Sound via the South Pole, but it soon became a race for survival amid the sea ice.

One night, however, the dangers of her fieldwork were brought into sharp relief. Alone on the sea ice, she heard a sudden 'boom':

My blood curdled, I felt cold down my back … and thought: 'OK, there's a crack opened up behind me.' I thought I'd get my equipment and move back—the crack had just opened, and I thought I'd keep walking just to make sure there weren't any more cracks. I kept walking and walking and the tide was going out. So I went back and packed up all my equipment, put it on my back and ran—by the end I was jumping floes! … The next morning I went out in the helicopter about 10.00 a.m. There was no pack there, it had all gone.[26]

The changeover

The arrival of a resupply vessel at Australia's Antarctic and subantarctic research stations has always been a special occasion. For some expeditioners, the 'changeover', as it was known, marked the beginning of their Antarctic experience, while for others it signalled the end of their Antarctic stay and their imminent return to routines and relationships back in Australia. The changeover was invariably a frenetic time, when all hands pitched in to unload equipment, construction materials and essential supplies. John Béchervaise described the 'utter weariness' of the changeover at Davis station during his voyage aboard the *Thala Dan* in 1959–1960, recalling how everyone on board was roused at 5.45am and taken to shore, where they would work until noon, returning to the ship for an hour before starting a second shift that would last until eight at night. The whole process took place over an exhausting five days and nights. Finally, those staying behind stood on the ice to farewell the ship, coming to terms with the fact that they would not see another vessel until their season came to a close.[27]

Technological advances and refined procedures have improved the efficiency and safety of the changeover since the early days of the Australian National Antarctic Research Expedition (ANARE), although the unpredictability of Antarctic conditions—fondly known as the 'A Factor'—still reigns supreme.[28] In November 2006, the *Aurora Australis* was

↓ All hands on deck for the changeover, 1993.

← Expedition members unloading stores from the ANARE relief ship *Kista Dan* during the changeover at Mawson station in 1955. Over three weeks, they offloaded over 450 tonnes of stores and undertook an extensive program to double the number of buildings at the new station.

struggling to approach Mawson station after delivering equipment to Macquarie Island and a group of summering expeditioners at Casey. It was soon clear, however, that the unusually heavy sea ice and volatile winds that season would prevent the ship from landing, at least in time to complete the final leg of the voyage to Davis before heading home. Suddenly, the weather momentarily cleared and senior personnel made the decision to conduct a changeover of sorts, deploying the ship's helicopters to transfer 15 people and essential cargo to Mawson and retrieve nine 'winterers' after their year working in one of the most isolated and extreme environments on Earth.[29]

Tracking emperors

Dr Barbara Wienecke was running on adrenaline after a wild voyage from Hobart in massive seas. It was 1994, and her mission as ANARE seabird ecologist was to study the emperor penguins at the Auster colony, near Mawson, over the winter months, continuing work begun by Graham Robertson in 1988.[30] Emperor penguins are the only Antarctic creature to breed during winter, the female hatching a single egg on the sea ice from mid-July. The male remains at the colony to look after the hatchling, while the female spends two months away foraging before returning to relieve her mate.

Edward Wilson had been intrigued by the emperors' curious breeding practice when he first saw fledged chicks on the sea ice at Cape Crozier in summer during Scott's *Discovery* expedition in 1901. Wilson speculated that the chicks must have hatched from eggs laid in midwinter, and he was determined to return to Antarctica during winter to retrieve live eggs for closer examination. At the time, emperor penguins were considered to be a primitive form of bird, and some thought it likely that their embryos would reveal the evolutionary link between reptiles and birds. Wilson did return to Antarctica, this time with Scott's *Terra Nova* expedition. In midwinter 1911, Wilson and two companions (Apsley Cherry-Garrard and Henry 'Birdie' Bowers) embarked on a 100 kilometre trek to the colony to retrieve five live embryos. Two eggs broke on the return trek, but Wilson managed to extract and preserve two of the remaining three embryos and donate them to the Natural History Museum in London. By the time the embryos were examined in 1934, however, scientists had rejected the evolutionary theory that inspired the expedition. As bizarre as it was, their

← Edward Wilson's 1903 watercolour sketch of 'Two emperor penguins, both carrying chicks (Cape Crozier rookery)' from Scott's *Discovery* expedition.

winter quest would become immortalised as one of the epic journeys of the era.[31] Eighty years later, Wienecke and her assistant Kieran Lawton embarked on their own midwinter journey to discover the mysteries of the emperor penguin. Their destination was the Auster colony, 51 kilometres from Mawson, and they would undertake their mission with the benefits of modern transportation and communications. Nevertheless, their fieldwork would demand no less courage and stamina than that first winter journey.[32]

The ocean was calm when the *Aurora Australis* finally entered the sea ice off Mawson, and Wienecke caught her first glimpse of the great continental ice sheet from the bridge: 'It's just such a phenomenally beautiful place—you're just not prepared for it. It really was so awesome and truly awe-inspiring.'[33] Nearer the harbour, she could see the big colourful buildings of Mawson station arranged like some giant 'LEGOland'. Even by the early 1990s, there were still few women working as scientists on the continent. That year, there were just herself and a female chef at Mawson, and Wienecke would be the first woman to spend a winter working out on the ice. Once the sea ice was thick enough, she and Lawton would begin their fieldwork, based in a tiny hut near the colony, attaching data loggers to individual adult penguins to track their foraging trips, and taking samples and measurements to determine their food consumption and physical condition.

By 17 May 1994, the sea ice was firm enough to take the quad bikes, and Wienecke and Lawton set out for their winter quarters. It was still early in the season, and the sea ice was susceptible to cracking, so two other expeditioners from the station accompanied them to test the stability of the ice by drilling a hole every 500 to 1,000 metres. Finally, they arrived at the little hut that would be their home for the next seven months.

The previous wintering group had added a cold porch to allow for outside clothes to be removed before entering the hut. Inside, she found four bunk beds, a few shelves, a gas stove and heater, a table and sink, and a 'beautiful big window'. The shelves, however, were stacked with nothing but cans of baked beans, and Wienecke hated baked beans: 'I'd made the mistake of mentioning this to the gentleman who had worked the previous winter there!' Fortunately, the pair had brought their own supplies, having hauled everything they needed on sleds behind their quad bikes. Fresh water was readily available from some icebergs nearby, and they would only need to travel to the station every few weeks to shower and restock the shelves. Otherwise, they were on their own.

Wieneke knew that the emperor penguin colony was located about 12 kilometres from the hut, but the colonies are highly mobile and, each winter, the birds spread out on the ice. The first task was to navigate a safe route for the quad bikes across the newly formed sea ice to the colony and, on that first day out, they had only a rough idea of where the colony might be.[34] After several kilometres of riding among grounded icebergs, they had seen just one bird. Every now and then they would stop to listen for calls, but the icebergs muffled all sound:

Then, at the end of this huge, tabular iceberg we had turned slightly right again and suddenly there they were … there was just this mass of black bodies and the noise was phenomenal.[35]

There were around 11,000 pairs of birds, most of which were engaged in courtship rituals and promenading with potential partners. Wienecke fell on her knees at the sight: 'Nothing prepares you for something like that'. When the sea ice was thick enough, they would travel to the colony with a Hägglunds all-terrain vehicle. As Lawton drove, Wienecke would sit in the back peering at the radar screen, learning to navigate to the colony by identifying individual icebergs in case they should ever get caught in

↑ Emperor penguins at Auster colony, approximately 55 kilometres from Mawson station. The precise location of the colony depends on icebergs and sea ice conditions in any given year.

white-out conditions. There were icebergs all around them, each with its own distinctive shape, so she perfected the art of identifying each one along the way on the radar screen and naming it on her mental map of the sea ice. The first iceberg out from the hut was the 'Pig', then came the 'Amphitheatre' and the 'Mother of all Bum Slides', and finally the 'Big Window'. The technique worked well and, on a couple of occasions, even saved the intrepid researchers from losing their way in white-out conditions.

The immediate task at hand was to establish a base near the colony in the form of a prefabricated Apple hut, sufficiently lightweight to enable easy deployment in the field. Wienecke chose a site on flat ice about 500 metres from the edge of the colony, then sat back to ponder how she would catch these enormous birds to study them. Her experience in studying little penguins (*Eudyptula minor*)—the smallest penguin species—on Penguin Island south of Perth had only partly prepared her for this. Here she would have to catch and attach a tag to a penguin that could weigh up to 40 kilograms: almost as heavy as herself. As with iceberg navigation, she perfected her technique, conscious of minimising handling to avoid stressing the bird. She would approach a suitable candidate, keeping low and moving slowly, careful to avoid looking at it directly:

> At some point you just have to take your heart in your hands and get the shepherd's crook around the neck. They tend to flop on their belly, so you hang onto the end of the crook and run with them, moving hand over hand to the far end of the crook then throw one arm in front while throwing the crook away with the other.[36]

She was careful to not hold the flippers down. An emperor's flippers are their main weapon of defence, and they are enormously powerful. Needless to say, her arms were soon covered in bruises from wrist to armpit, so she improvised by making arm shields from pieces of an old foam mattress and thick socks. One big girl managed to knock her unconscious. The penguin had seemed calm enough, then she lifted her right flipper quite suddenly and hit Wienecke across the nose: 'I saw stars'. This happened again when the same penguin returned a couple of months later, this time knocking out one

❯ ANARE seabird ecologist, Dr Barbara Wienecke, undertaking fieldwork at the emperor penguin colony near Mawson station in 2006.

of the tradies from the station while he was helping Wienecke remove her tag. The other emperors seemed to find such antics entertaining. Each morning a group of about 20 or so non-breeding birds would peel away from the colony and gather at the hut, ready to follow the visitors around all day.

Wienecke and Lawton managed to tag 12 females and three males during that winter. It was the first time in ANARE history that the penguins' movements were tracked via satellite for the duration of their foraging trips. Only one of the females failed to return, her tracking device indicating that she had come to within 30 kilometres of the colony, then apparently joined a group of males heading north. The tracking devices on the female birds showed that, although they were never more than 200 kilometres away from the colony, they would cover up to 3,000 kilometres in their hunt for food in the ocean before returning to the colony, fat and glossy, to reunite with their partner and take over the care of their chick. The males, meanwhile, stayed behind to incubate the egg and nurse the chick. During the long months of winter blizzards and darkness, Wienecke observed the males huddled together for protection and warmth, all the while balancing their precious single egg on top of their feet. Occasionally one would shuffle away from the group to eat snow, the egg pressed hard against its belly.

By the time the females returned in July, the males, who had not eaten for around 115 days, had lost almost half of their total body mass. Wienecke described how, when the females began returning, the normally subdued colony would erupt in an explosion of noise as the females called to locate their mates, producing two sounds simultaneously in their syrinx and effectively singing a duet by themselves. The males, she observed, were often reluctant to hand their chick over, and the females would have to bully them into relinquishing their offspring. Once the changeover occurred, however, the males would roll in the snow with 'flippers and feet everywhere', before heading off across the sea ice. They would trek some 70 or 80 kilometres to open water to have their first meal in more than three months, before returning to the colony to take their turn to feed the chick.

> Emperor penguins are uniquely adapted to survive in Antarctica's extreme conditions. Adults stand at around one metre tall and are among the largest of all living birds.

An ocean of riches

The Southern Ocean, renowned for its tempestuous westerly winds and currents, hosts an extraordinary abundance of marine life, thanks to the great upwellings of deep ocean waters south of the Antarctic Convergence. These upwellings create massive blooms of microscopic phytoplankton in the cold polar waters, attracting vast swarms of protein-rich Antarctic krill. Marine predators—baleen whales, seals, penguins and seabirds—converge in large numbers to feed on these shrimp-like crustaceans that grow to about the size of a human thumb and are one of the most abundant animal species on Earth.[37] When global fishing fleets began extending their operations into the Southern Ocean in the 1960s, alarm bells rang for Antarctic scientists, who understood that the little-known organism was at the heart of the Southern Ocean ecosystem and that overfishing would be catastrophic for the creatures that depended on them, particularly seals and whales.

Their efforts to focus attention on the role of krill in the Southern Ocean led to an ambitious ten-year international research program known as Biological Investigations on Marine Antarctic Systems and Stocks (BIOMASS). The program involved 15 ships from 11 Antarctic Treaty member nations, including Australia. The urgency posed by unregulated krill fishing was also the focus of discussions among Treaty parties, leading to the Convention on the Conservation of Antarctic Marine Living Resources (CCAMLR), which came into force in 1982. The Convention extended the reach of international governance under the Antarctic Treaty to include all marine living resources as far north as the Antarctic Convergence, around latitude 55° S. It recognised that predators and prey formed part of an interconnected marine ecosystem, and that unregulated krill fishing—or even a poor breeding season—could have a detrimental impact on all species that depended on it for food, and especially on those with no alternatives, such as the great baleen whales.[38] Australia hosts the secretariat for CCAMLR, and member nations meet each year in Hobart, Tasmania.

→ With the return of 24-hour sunlight, the Ross Sea bursts with life as phytoplankton blooms provide a banquet for krill, fish, penguins, whales and other marine creatures of the Southern Ocean.

Frostbitten

'The weather,' as Wienecke recalled, 'was foul, it was disgusting. Mawson is south of the Antarctic Circle, so we lost the daylight for a couple of weeks.' The researchers would get up at 10am and, with the aurora raging overhead, travel to the Apple hut in the darkness so as to spend as much time as possible observing the colony. During one such excursion in early July, they both stopped in their tracks as 'an absolutely extraordinary ruby-red glow' suddenly appeared ahead of them between two icebergs. It took them a moment to work out what they were looking at. The sky had been the same dull grey for weeks on end, and the sun had momentarily appeared above the horizon before sinking again.

Sometimes, blizzards forced them to seek shelter in the Apple hut. On one occasion, Wienecke and Lawton were out collecting abandoned eggs before a blizzard set in out on the sea ice. Wienecke recalls that she lost radio contact with her companion and found herself close to an iceberg in deep snow. As the wind picked up, her goggles became useless and she pulled them off, only to have icy wind biting into her eyeballs. If she had blinked, her eyelids would have frozen shut, so she turned her head and tried to peel the ice off her eyelashes. The thought occurred to her that she might die, but then:

> While I was sitting there, out of the blowing snow came a single penguin, tobogganing on its belly gracefully, effortlessly past me. It was just about an arm's length away from me and this bird just looked at me as it was passing and I'm sure it thought 'you fool, what are you doing here?' and off he went to join the colony. It was a magic moment.[39]

→ Cecil Madigan, meteorologist with
the AAE, sporting an ice mask after
working outside during a blizzard.

↙ An emperor penguin at Mawson
ice edge.

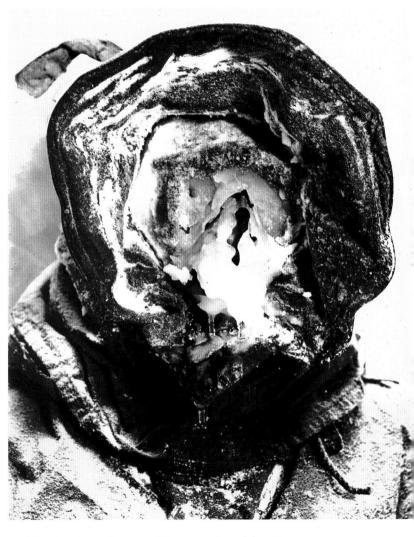

The chance encounter gave her the strength to fight on, and she managed to reach the Hägglunds just as the full-blown blizzard arrived.

Sometimes a blizzard would keep them confined to the main hut for days, and Wienecke became proficient at shovelling the snowdrift that would build up and compact hard against the hut entrance. She remembers being cold much of the time, but she had no regrets. This was her passion and, besides, 'you're there to do a job and survive'. Overnight, when the temperature inside the hut plummeted to −26°C, she would reflect on the conditions endured by Mawson and his men. She particularly sympathised with their frostbitten fingers. Her own were permanently damaged by frostbite but, for all the hardships, she revelled in the isolation and having the opportunity to witness the incredible tenacity and dedication of the emperor penguins. 'In that moment', she recalled, 'I would not have wanted to be anywhere else on this planet but there.'[40]

So this is an experience, then, that you don't usually have in your familiar surroundings, where your fear and your wonder are mixed up and it's difficult to separate them.

SIR SIDNEY NOLAN[1]

Wilderness

← McMurdo Sound, at the edge of the Ross Ice Shelf, was discovered in 1841 by the English explorer James Clark Ross. It became one of the main access routes to the Antarctic continent and the location of scientific research stations operated by the United States and New Zealand.

For over two centuries, Antarctica has enticed Australian explorers, writers, photographers, visual artists, musicians, adventurers and tourists. For many, it is simply a white space on a world map, close yet unreachable, a vast ice canvas stretching around the pole, etched with stories and legends.[2] Most of those who have visited long enough to feel its chill air and embrace its silences have been appointed by the Australian government to undertake exploration, science or station construction and maintenance. A few have ventured there on more personal quests: to absorb its otherworldliness, to refresh the spirit, or to test the limits of their physical and mental endurance.

→ Antarctica's isolation and otherworldly qualities have long stirred the imagination of adventurers and artists.

Learning by doing

In the summer of 1987, three visual artists embarked on a voyage to Antarctica with Australian National Antarctic Research Expeditioners aboard the *Icebird*. Bea Maddock, Jan Senbergs and John Caldwell were offered this rare chance to accompany a group of scientists and others on board an annual resupply voyage to Australia's Antarctic bases as part of the Australian government's new Humanities Berths Program. For the three artists, the voyage was an attempt to draw Antarctica into Australia's cultural consciousness.[3] Barry Jones, the Australian government minister responsible for science

and Antarctic affairs, was keen to demonstrate Antarctica's potential as a source of inspiration to all Australians. A year earlier, he had received some unwelcome news from his friend Stephen Murray-Smith, who had visited Casey and Davis stations to observe the cultural life of Australia's Antarctic stations during a period of changing attitudes and values in Australian society.

Murray-Smith found an Australian Antarctica that had a strong frontier ethos, but with little sense of history or tradition. 'Each year here,' he observed, 'begins anew with the breakup of the winter ice.'[4] It seemed to him that Australians were given little opportunity to learn about Antarctica, and he blamed everything from Australia's school curricula to the lack of annual expedition records kept on the bases or a decent library that expeditioners could consult. New expeditioners would arrive without any understanding of how their work was part of some larger project or how past experiences of others might offer valuable insights or lessons. There was an irony, he observed wryly, in the Australian government sending personnel and supplies year after year and engaging in a major building program designed to consolidate Australia's presence in Antarctica for decades to come, but allowing the knowledge and experience so painstakingly gained to dissipate at the end of each voyage along with the expeditioners, many of whom would never return to the icy continent.[5] By failing to attend to the social environment of the stations, he argued, the Antarctic Division planners had unwittingly created a culture where regulations mattered more than 'learning by doing', resulting in little shared understanding of why Australia was there at all.[6]

For Murray-Smith, this lack of a sense of history reflected a wider ignorance among Australians about their country's long association with Antarctica. While generations of Australian school children still embraced the heroic stories of early explorers such as Scott, Shackleton and Mawson, the nature of the region and Australia's presence in it seemed to have been lost on them. Even successive Australian governments seemed oblivious to the events and practices that had given impetus to their own policies and programs. 'Surely,' he wrote wearily, 'we can't understand the tasks that lie before us without knowing what they are the end product of?'[7] Sending artists and writers to Australia's Antarctic bases seemed to offer one way to bring Antarctica to Australians, most of whom would never experience the continent firsthand, so that they might gain some appreciation of what lay beyond the heroic stories and white blankness of the ice sheet. Through their imaginative responses to the Antarctic environment, they could stimulate a deeper interest in the study of Antarctic history and culture.[8]

Picturing Antarctica

The three artists who made their way south in 1987 were already immersed in the history of elaborate medieval maps of the fabled southern land produced by early cartographers, and imaginative renderings of strange creatures that might inhabit the far southern extremes of the globe.[9] Official artists accompanying the earliest voyages of exploration

↓ This 1650 map, *Polus Antarcticus*, showed the extent of geographical knowledge of the southern polar region in the 17th century.

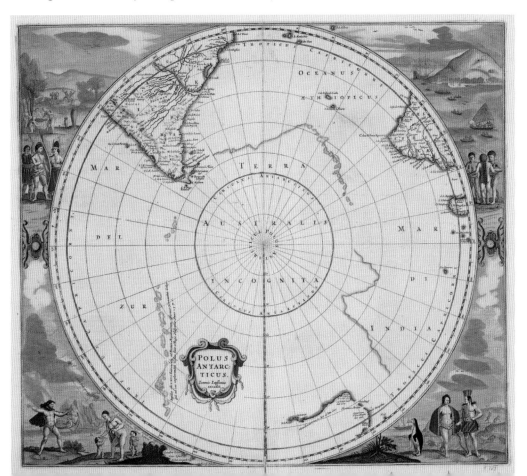

in the southern polar region had created visual records of the landscapes, geology, flora, fauna and people encountered during the journey, since James Cook engaged William Hodges as the official artist on his second voyage of maritime exploration in 1772–1775. But while watercolours and drawings were easily rendered aboard a moving vessel, it was the medium of photography that would come to dominate visual representations of Antarctica from the early twentieth century.[10]

Australians figured prominently in this reimagining of the mysterious neighbouring continent. Among them was Frank Hurley, a photographer from Sydney, whose name—like Mawson's—would be inextricably woven into the fabric of Australian Antarctic history.[11] Hurley had already acquired something of a reputation for his technical camera skills when Mawson invited him to accompany the Australasian Antarctic Expedition (AAE) as official photographer. With training and support provided by the film production house Gaumont Company Ltd, Hurley found himself on the *Aurora* bound for Antarctica in December 1911 with some of the most advanced equipment of the day, including a variety of still cameras and a cinematograph camera for moving images.[12] Hurley's enthusiasm for the job knew no bounds. He rarely went anywhere without a camera, and climbed into the ship's rigging to capture the ship pushing through pack ice. His antics elicited admiration from his fellow expeditioners. As the expedition's doctor, Archie McLean, observed:

> *One of his earliest escapades was to climb up into the crow's nest—on the main-mast top, some sixty feet [18 metres] from the deck while the ship was heeling over and pitching in her most formidable style ... accompanied by his 'old friend'—a reflex half-plate camera which secured for him subsequently many exquisite photographs.*[13]

Mawson's account of the expedition, published in 1915 as *The Home of the Blizzard*, included many of Hurley's black-and-white photographs, although Hurley's untimely departure for an outback adventure with Francis Birtles meant that his autochrome colour plates were missing from the publication.[14]

→ Frank Hurley working on the tip of the jib-boom of the *Discovery*, in 1930. Hurley's images brought the Antarctic region to life for Australian audiences, capturing its beauty and wildlife as well as the challenging conditions of early exploration.

→ One of the most iconic images of the 'heroic era' of Antarctic exploration is Frank Hurley's 1915 photograph of the *Endurance* being slowly crushed by sea ice.

← Frank Hurley was admired for his ability to take photographs under arduous conditions.

Hurley's polar adventure with Ernest Shackleton's *Endurance* expedition in 1914–1917 established him as a household name, a popularity that he relished over the coming decades. The captain of the *Endurance*, Frank Worsley, was himself an admirer of Hurley's skill and perseverance as the expedition photographer. He wrote:

> H is a marvel—with cheerful Australian profanity alone aloft & everywhere, in the most dangerous & slippery places … content & happy at all times but cursily [cursing] so if only he can get a good or novel picture. Stands bare & hair waving in the wind, where we are gloved & helmeted, he snaps his snaps or winds his handle turning out curses of delight & pictures of Life by the fathom.[15]

When the *Endurance* became beset by pack ice and was eventually crushed by it, Hurley recorded the ship's death throes in a series of images that would become one of the most evocative renderings of Antarctic exploration of all time. While the ship slowly succumbed to the ice, Hurley, with the help of a seaman, managed to dive under a metre or more of mushy ice to rescue his boxes of precious negatives and cinefilm from the compartment that had served as his darkroom store aboard the ship. He kept the cinefilm and chose 150 of the best glass plates, smashing another 400 to reduce weight for the sledging journey that followed.[16] The value of these surviving negatives and film is now inestimable.

↓ 'The landing on Elephant Island, solid rock lies beneath our feet, this was Paradise regained!' *Frank Hurley, 15 April 1916.*

So began one of the most epic journeys in Antarctica. Over the next few months, the men hauled three lifeboats across floating pack ice before reaching land on the tiny outcrop of Elephant Island. While Shackleton and five others embarked on their own remarkable voyage into the stormy Southern Ocean to seek help from whalers on South Georgia island, Hurley and 21 others awaited rescue for 18 long weeks, sheltering beneath two upturned lifeboats and surviving on a meagre diet of penguin, seal and lichen. Hurley was deeply affected by the experience, writing a poem to convey its impact on him and his companions:

We had suffered, starved, and triumphed, grovelled down, yet grasped at glory,

We had grown bigger in the bigness of the whole.

We had seen God in His splendours, heard the text that Nature renders

We had reached the naked soul of man.[17]

→ Hurley photographed the stranded Elephant Island party two weeks after Shackleton's departure in search of help. Frank Wild is seated second from left. The men endured a brutal winter before being rescued. Shackleton later remarked: 'Not a life lost, and we have been through Hell!'

Hurley marvelled at how, even in their dire circumstances, they managed to celebrate midwinter. 'I wonder if a popular concert,' he recalled, 'was ever conducted under more peculiar conditions than that midwinter revel of ours.' He continued:

> *For the concert-hall is but four feet nine inches [1.45 metres] high, and for an assemblage of twenty-two provides only lying-down accommodation ... The programme of that midwinter concert may not have been high art, but Covent Garden has held no more appreciative audience ... Of the thirty-odd items, fully half were topical songs, stories and recitations, on which the brains of members had worked overtime for days, and no body of undergraduates ever relished their own wit more keenly, or roared their topical choruses with greater fervour.*[18]

↓ **The distinctive cone-shaped rock off Elephant Island where Frank Hurley and his companions survived for four months awaiting rescue.**

↑ Frank Hurley's impression of the makeshift sleeping quarters they called The Snuggery.

↑↑ 'Our home on Elephant Island was built of two upturned boats laid side by side, twenty-two of us lived like semi-frozen sardines within its cramped, dark interior.' *Frank Hurley, 1916.*

Mawson's Huts

The centenary of the AAE in 2012 served as the focus of several pilgrimages to the Antarctic continent, including the Australian Antarctic Division's voyage to Mawson's Huts in Commonwealth Bay. By then, Douglas Mawson's name had become synonymous with Australia's national pride in its Antarctic history, and the group of buildings established by Mawson as the Main Base for his AAE in 1911–1914 was one of the most widely recognised places associated with Australia's early history of Antarctic exploration.[19] The site, including the Main Base Hut, magnetograph hut and transit hut (astronomical observatory), is notoriously difficult to access because of the strong katabatic winds for which Commonwealth Bay became renowned. Nevertheless, the heritage significance of Mawson's Huts is recognised under the Antarctic Treaty and continues to attract visitors seeking to reconnect with the tangible legacy of Australia's contribution to the 'heroic' era of exploration. The Australian government added the Mawson's Huts Historic Site to the Commonwealth Heritage List in 2004 in recognition

↑ Mawson's Huts are of national heritage significance for their association with Australia's role in Antarctic exploration and science in the early 20th century

↑ Mawson's Huts are one of only six expedition bases
to survive from the 'heroic era' of Antarctic exploration.

of its significance as one of only six remaining wintering expedition bases built in
Antarctica during that early period of Antarctic exploration. It was also Australia's first
base for scientific and geographical discovery on the continent. While most Australians
are unable to visit the site, the Mawson's Huts Foundation has built a replica of
Mawson's Main Base Hut on the Hobart waterfront—200 metres from where Mawson's
expedition departed in 1911—to give visitors a taste of the living conditions experienced
by these early expeditioners.[20]

← Several members of the AAE, including the captain of the *Aurora*, published their own accounts of the expedition.

Showcasing Antarctica

From the start, those returning from Antarctica discovered that Australians had a passion for stories and memorabilia from the far south. When Shackleton's *Nimrod* expedition berthed in Sydney in 1909, the ship attracted tens of thousands of visitors hungry for a souvenir of rock, animal skin, eggs or even a piece of the vessel itself.[21] Many early expeditioners became enthusiastic promoters of their own achievements, already mindful of their place in the history of polar exploration. The importance of self-promotion was incalculable in an era when Antarctic expeditions depended upon private investors and financial returns, with expeditioners producing books, lectures, lantern-slide shows, exhibitions and films. When Douglas Mawson—who accompanied Shackleton as the expedition geologist—gave a lecture on his southern polar experiences to a crowded audience at the Prince of Wales Theatre in Adelaide in July 1910, he stepped onto the stage dressed in his full Antarctic regalia. As one observer wrote:

> *The lecturer looked tremendous in front. He forthwith removed several pairs of socks from the wind and snow proof Burberry jacket. Having taken off the rubber-like coat he removed an arctic jacket of pure wool weighing 14 lb [6.5 kilograms]. Underneath again was another warm jacket, but he did not divest himself of any more clothing, and still he was perspiring profusely. On his hands were arctic wolf gloves and underneath fingerless mittens, and on his feet he wore reindeer-skin boots. Having completed the story of arctic expeditions down to the days of Peary, and illustrated his remarks with informing maps and lantern views, Dr. Mawson told the history of antarctic expedition up to and including the Shackleton expedition.*[22]

In the same year, an exhibition opened in Adelaide featuring the skis, sledge and tent poles used by Mawson. The exhibition attracted nearly 200,000 visitors over a six-week period.[23]

Hurley, too, was encouraged by the Australian public's thirst for stories and images of Antarctic exploration. On his return from the AAE with Mawson in 1911–1914, Hurley used his cinematographic footage to piece together a film of the conditions in Antarctica as the men went about their daily lives. Mawson understood the value of such imagery in attracting investors for future expeditions, but his primary interest still lay in disseminating the scientific results of his expeditions. In a letter to Hurley in 1916, he wrote:

> Al[l] expeditions are in the public eye at the moment of return but afterwards it is the published scientific matter (and photographs are such) upon which are based estimates of the expeditions.[24]

Following the extraordinary survival and rescue of the *Endurance* expeditioners from Elephant Island, Hurley began touring with his expedition films and photographs, marking the beginning of his career as one of the most celebrated showmen and chroniclers of the early twentieth century.[25] When he returned from Mawson's British, Australian, New Zealand Antarctic Research Expeditions (BANZARE) in 1930, Hurley produced another two films: *Southward Ho with Mawson* and *Siege of the South*, and toured extensively with what he called his 'synchronized lecture entertainments' to wide acclaim and packed halls in cities and country towns around Australia, cultivating his own legendary status as an Antarctic pioneer in the process.[26] Hurley's photographs and movies were not only powerful visual displays of the Antarctic environment. They also served to consolidate Antarctica in the settler Australian imagination as a new kind of colonial frontier, where polar explorers like Mawson became folk heroes for a new nation keen to prove its worth.[27]

→ Most Australians would have been unfamiliar with the type of gear required for Antarctic conditions.

Protecting Antarctica and the Southern Ocean

With the signing of the Antarctic Treaty in 1959, a unique international consensus emerged from a group of 12 countries whose scientists had been active in the Antarctic region during the International Geophysical Year in 1957–1958.[28] The signatories agreed that the land and seas south of latitude 60° S would be dedicated to peace, science, conservation and international collaboration. One of the 14 articles provided that any party could inspect the ships, stations and equipment of other parties in Antarctica to ensure that they were compliant with the treaty.[29] Seven of the signatories—including Australia—had existing sovereignty claims in Antarctica and, while some countries did not recognise the claims, the treaty explicitly preserved their 'status quo' so that no activities could take place changing their status and no new claims could be made during the life of the treaty.

A series of related international agreements followed, forming part of the Antarctic Treaty System and providing for regular consultative meetings of the parties, addressing matters such as scientific and logistical cooperation; conserving of environments, plants, animals and historic sites; designating protected areas; managing tourism; and collecting data and exchanging information. Specific conventions agreed by treaty

↓ Since the late 1960s, commercial tourism has given many Australians the opportunity to experience Antarctica for themselves, although visitors must adhere to strict guidelines to avoid damaging this sensitive environment.

parties include the Conservation of Antarctic Fauna and Flora (part of the treaty signed in 1959), subsequently replaced by the Convention for the Conservation of Antarctic Seals (CCAS, 1972), and the Convention on the Conservation of Antarctic Marine Living Resources (CCAMLR, 1980). The focus shifted from sea to land in the 1980s, as tensions heightened over the exploitation of Antarctica's potential mineral resources. The resolution of this conflict would take another decade—resulting in the Protocol on Environmental Protection to the Antarctic Treaty (the Madrid Protocol, 1991)—but it illustrated a major shift in public attitudes towards protecting and valuing the Antarctic environment.

The Antarctic Treaty System has proven to be a remarkably effective model of international governance, although territorial and sovereignty interests and pressures to exploit the region's resources are never far from the surface. In recent years, environmental advocates have lobbied for a network of marine parks to be established across the Southern Ocean. In 2009, CCAMLR members agreed to create the first marine park around the South Orkney Islands to address the increasing incidence of illegal and unregulated fishing of the Patagonian toothfish (*Dissostichus eleginoides*). They subsequently agreed to create the world's largest protected area, the Ross Sea marine park, in 2016. Nevertheless, the fragile international consensus reflected in the treaty and associated agreements will continue to be tested as the impacts of climate change and population increases are felt in this remote and unique polar environment at the southernmost end of the planet.[30]

The last frontier on earth

During the course of the twentieth century, stories of Australia's 'heroic' era of Antarctic exploration became a powerful cultural touchstone for successive generations of Australian artists and writers. The artist Sidney Nolan—best known for his paintings depicting Australian bush folklore—made his own pilgrimage to the polar continent in 1964, spending eight days at McMurdo base as a guest of the United States government during Operation Deep Freeze. Like many children of his generation, Nolan had developed a childhood affection for the romantic stories of Antarctic exploration. When his friend, the expatriate Australian writer and war correspondent Alan Moorehead, suggested the idea of an Antarctic trip, Nolan needed little persuading. Once out on the ice, he worked feverishly, sketching 200 watercolours in the field and subsequently using these (and his own photographs) to produce a series of 68 paintings back in his London studio.[31]

Nolan, who had long been preoccupied with the artistic potential of Australia's deserts, found something both alien and strangely familiar in the Antarctic landscape. In Nolan's hands, Antarctica emerged as a white desert, reminiscent of the red desert country of Australia as portrayed in his Burke and Wills and Ned Kelly series, painted around the same period. In each series, Nolan was interested in heroic but flawed explorers in a vast and hostile desert environment. Like Hurley before him, Nolan had a sense that there was something about the continent's extraordinary landscape that challenged the human imagination in a way that no other continent could. Reflecting on his Antarctic experience, he recalled that he had expected to see a 'flat, enormous paddock' down there, but instead 'found a majestic kind of great continent' that both haunted and exhilarated him:

> One felt this instantaneous fear at the first sight of it, that it would annihilate one; but this was overcome straight away by the sense of wonder in it ... It represented a reality stronger than oneself, and in a paradoxical kind of way one felt safe.[32]

→ For the Australian artist Sidney Nolan, the Antarctic landscape was both alien and sublime.

Writers, too, were drawn to the imaginative possibilities of the southern polar region. One of the earliest literary works in English to imagine the Antarctic region was Samuel Taylor Coleridge's gothic poem, 'The Rime of the Ancient Mariner', published in 1798. It appeared just 25 years after Cook had sailed among the 'ice islands', and it manifested many of the themes that still inspire Western art and literature about the Antarctic region: experiencing the voyage to the southernmost end of Earth as a 'journey of transformation' and discovering a hostile realm where humans are confronted by their own mortality in this world of tempestuous ocean and ice.[33] Such fictional accounts of the Antarctic world gained popularity during the nineteenth century, fuelled by reports from scientific expeditions returning from the far south. Some portrayed the continent as a malignant place where shapeshifting alien creatures emerged from the thawing ice.

Australian writers, too, were intrigued by Antarctica's unearthly geography and drawn to its imaginative possibilities. When Thomas Keneally first visited the continent in 1968 as a member of an official group led by the American ambassador to Australia, he was struck by 'the giant landscapes and improbable, barely polluted vistas of Antarctica in so profound a way that it recurred in my dreams for decades'. He wrote:

> It had provoked no native tongue, no rites, no art, no jingoism. Its landscapes existed without the permission of humanity. And everything I looked at, even the nullity of the pole, produced jolts of insomniac chemicals in my system. It was not landscape, it was not light. It was super-landscape, super-light, and it would not let you sleep.[34]

He visited again in 2013, ostensibly to return a portion of biscuit he had taken from Scott's hut at Cape Evans, but realised that his visit was more about seeing the icy landscapes once again with 'their air of calm self-absorption ... as they fill your sight like an independent and immaculate planet'. 'After Antarctica,' he reflected, 'nothing is the same'. Keneally's experience seemed to echo that of other writers for whom the voyage south offered not just a physical encounter with a hostile environment, but an inner journey of personal reflection and spiritual transformation.[35]

Destination Antarctica

During the first half of the twentieth century, few Australians, apart from sailors, scientists and the occasional bureaucrat, had any opportunities to experience the Antarctic environment firsthand. That changed in the 1960s when entrepreneurs, recognising the potential for Antarctic tourism, began offering privately funded cruises and charter flights. One of the earliest and more unusual Australian 'expeditions' took place on 13 February 1977 when Dick Smith, an electronics retailer, aviator, adventurer and self-made millionaire, organised a Qantas charter flight to the South Pole to celebrate the 50th anniversary of Sir Hubert Wilkins's pioneering Antarctic flight in 1928.

It was to be an expedition with a distinctively Australian flavour. Smith planned to overfly Macquarie Island and Commonwealth Bay during the course of a day, giving his 300 passengers an opportunity to imagine that they were retracing Mawson's AAE. At Sydney Airport, the departure board simply read: 'Qantas 144 D.S. Ant. Flt.—Sth. Mag. Pole'.[36] Among the passengers were Mawson's daughters, Pat and Jessica, and his grandson Gareth Thomas, who had been largely unaware of his grandfather's achievements while growing up.[37] After eight hours of flying, the moment came when passengers could gaze down upon the vast ice sheet below. It was a distant yet strangely tantalising encounter for all on board. As one journalist put it:

> To fly above the home of the blizzard was not to feel the sting of the wind and snow, nor the isolation and suffering of the explorers. A steward served champagne. We headed north to the summer.[38]

When Antarctic tourism became more established in the 1980s, Australians were as keen as any to set foot on the ice continent. A few hardy souls saw it as an opportunity to test their own physical and mental endurance in one of the world's last wildernesses.[39] Among them was an Australian couple, Don and Margie McIntyre, who lived for a year in a tiny, prefabricated hut anchored by chains to a rock near the site of Mawson's Huts. Sponsored by the *Australian Geographic* magazine, their stay in Antarctica was intended to showcase the physical and emotional conditions experienced by Mawson and his men, albeit with the benefits of modern equipment and communications

← Australian adventurers Justin Jones (left) and James Castrission reached the South Pole on 31 December 2011 during the first unassisted trek ever completed from the Antarctic coast to the South Pole and back.

→ A group of Antarctic tourists prepare to spend the night in tents on the ice. Antarctic tourism operations are governed by the Antarctic Treaty System.

technology.[40] In 2007, another adventurer sought to follow in Mawson's footsteps, this time by re-enacting Mawson's 500 kilometre trek on the Far Eastern Party's sledging journey. Tim Jarvis, a British–Australian scientist, wanted to replicate the experience as authentically as possible by choosing equipment, rations and clothing like those used by Mawson in 1912. Dogs were no longer allowed on the continent, so Jarvis made do with a heavy wooden sledge, reindeer-skin sleeping bags and a tent without a floor.[41] He later wrote how, in the midst of an Antarctic night, he could well imagine Mawson's state of mind as he lay starving and alone on the plateau. 'On polar explorations,' Jarvis wrote, 'you play all sorts of mental tricks to keep yourself going, operating for much of the time in a very altered reality.'[42] Four years later, James Castrission and Justin Jones completed the first unassisted traverse on foot from the Antarctic coast to the South Pole and back, covering 2,300 kilometres in 89 days.[43]

Other adventurers have been drawn to the continent or islands to swim or dive or kayak in ice-filled waters or to climb mountain peaks, while a steady stream of the less adventurous cruise south each summer season to experience the polar environment firsthand. Antarctica, long considered too remote and inaccessible for the casual visitor, has become a destination of choice for those who can afford it. As the demand for Antarctic tourism has increased, however, so too have the inherent safety risks and potential environmental impacts of private expeditions and tourist vessels venturing into this fragile region.[44]

So far, the vast natural scientific laboratory that is Antarctica has been preserved in pristine form. History will show whether it will remain that way or whether the depredations of man, driven by his urgent need for mineral and other resources, will pollute and contaminate its environment and disrupt its fragile ecological balances.

PHILLIP LAW[1]

Environment

Antarctica in a warming world

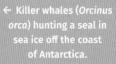

← Killer whales (*Orcinus orca*) hunting a seal in sea ice off the coast of Antarctica.

When British polar scientist Dr John Heap addressed the ANARE Jubilee Science Symposium in 1997, he observed that Antarctic science was at a 'crossroads'. There was an urgent need to understand the changes occurring in the Antarctic ice sheet and their implications for a warming planet:

You are inheritors of Douglas Mawson's scientific mantle … You have a huge area of Antarctica, both land and sea, at your back door. There are immense scientific opportunities and challenges in the Antarctic, but so also are there in Australia. Choices have to be made … there are some who do not wish to know, for example, about the strength or weakness of the thread from which hangs the Sword of Damocles represented by the Antarctic ice sheet.[2]

By the end of the twentieth century, the era of the 'individual adventurer–scientist' championed by Douglas Mawson and Phillip Law was over. It had been overtaken by international scientific collaborations seeking answers to complex new questions about the Antarctic environment. The challenge for the new century was not simply to determine the nature of the southern polar region, but to understand what the region could reveal about global climate change. The depths of the Antarctic ice sheet and the surrounding ocean had become Australia's new frontiers.[3]

The resource era

For much of the twentieth century, enthusiasm for Antarctic exploration and science was piqued by the possibility of discovering a new source of minerals beneath the massive continental ice sheet.[4] As a geologist, Mawson paid close attention to areas of exposed rock in the ice sheet, this 'natural museum' where the

← As a geologist, Douglas Mawson was intrigued by the 'natural museum' of ancient rock pushed to the surface of the ice sheet, revealing the promise of the mineral wealth that lay beneath.

↑ A geologist at the US Antarctic research station Little America examines core samples taken from the Bay of Whales in 1958. By the end of the 20th century, ice core research had become an international effort with scientists from different countries collaborating to unlock the secrets of Earth's climate history from deep within the Antarctic ice sheet.

movement of glacial ice had plucked boulders from the bedrock and pushed them to the surface or pulverised them into rock flour.[5] Apart from its 'great antiquity', Mawson was intrigued by tiny veins of quartz carrying small quantities of less common minerals, such as beryl, tourmaline and garnet, as well as the ores of iron, copper and molybdenum, indicating the 'probability of mineral wealth beneath the continental ice-cap'.[6] The promise of Antarctica's mineral wealth resonated powerfully in Australia. In May 1939, as war loomed in Europe, the *Kalgoorlie Miner* speculated on the vast riches awaiting discovery within 'Australia's Antarctica': 'It is almost certain that one day man will find some means whereby he can tap at will the rich resources of the frozen continent'.[7]

By the postwar decades, the stakes were high for a world hungry for new natural resources. The idea of mining beneath the ice sheet may seem alien from a modern environmental viewpoint, but it represented the kind of 'utilitarian' conservation ethic that Mawson had championed in the 1920s and 1930s and Phillip Law was advocating in the 1960s. When the United States research ship *Glomar Challenger* found traces of hydrocarbons in the Ross Sea in the early 1970s, all eyes turned to the Antarctic region.[8] The discovery by Soviet reconnaissance parties of iron ore deposits in the region in 1976 fuelled predictions in Australia that a 'mountain' of iron ore would be found in the Prince Charles Mountains within the Australian Antarctic Territory.[9] The 'scientific era' that had defined the International Geophysical Year (IGY) of 1957–1958 in Antarctica seemed to be transforming into a 'resource era'.

The world was warming up for a new kind of international race to the southern polar continent. The problem for member nations of the Antarctic Treaty, however, was that the treaty did not explicitly address the region's potential mineral resources, and unregulated mining and oil exploration would have serious implications for both the environment and international relations in the region. By the 1970s, the treaty nations were under increasing pressure to impose a moratorium on mining in Antarctica. In 1981, international tensions were high as they began negotiations to seek a consensus on how to regulate mineral exploration and develop a comprehensive minerals regime in Antarctica. By then, the environmental impacts of the human occupation of Antarctica were under increasing public scrutiny. While the treaty nations were grappling with the question of Antarctica's potential mineral resources, Greenpeace joined forces with other environmental organisations to campaign for Antarctica to be declared a World Park. Dr Geoff Moseley, then Director of the Australian Conservation Foundation, summarised the environmentalists' argument against mining in Antarctica:

> It's an area that is a great repository of knowledge about the past. It's a remote place for monitoring the environment and plays a role in climate. But its greatest value is as a wilderness area and for its educational and scientific values. It could be an example of restraint; that you don't have to use mineral resources just because they are there.[10]

In 1986, Greenpeace ships began a series of annual summer voyages to Antarctica to inspect government activities on the continent and document evidence of environmental devastation.[11] In order to argue its case under the Antarctic Treaty, however, Greenpeace decided that it needed to establish a permanent base there and, in 1987, the *Greenpeace* set sail for Antarctica with a group of volunteers who would establish the first permanent non-government base on the ice. From the World Park Antarctic base at Cape Evans on Ross Island, Greenpeace volunteers managed to maintain a presence on the ice for four years, monitoring pollution and construction activities on neighbouring stations. Meanwhile, other environmental activists joined forces with Greenpeace under the Antarctic and Southern Ocean Coalition to campaign against the signing of the Convention for the Regulation of Antarctic Mineral Resource Activities (CRAMRA) and advocating instead that Antarctica be declared a World Park where mining was banned.[12] By June 1988, however, the treaty nations were ready to sign the CRAMRA after seven years of negotiation.

At the time, Australia was engaged in its own public debate about CRAMRA, with the mining industry and conservation groups at odds over whether the Australian government should sign on to the convention at all, and the Federal Cabinet also divided on the issue. The French government had similar reservations and joined with Australia in refusing to sign it. Over the next 18 months, Australian diplomats led attempts to create a consensus for a new environmental regime that excluded mining.[13] Two catastrophic events in 1989 helped to galvanise public opinion in favour of protecting Antarctica. In January, an Argentinian naval ship loaded with oil was wrecked in Antarctic waters near Anvers Island, releasing a thick black slick along three kilometres of coastline. *The Sydney Morning Herald* reported that '[p]enguins are emerging from the icy depths covered in oil', their oil glands clogged, while other animals such as seals would be affected by the destruction of Antarctic krill. A month later, the *Exxon Valdez* ran aground in Alaska, creating the world's most environmentally destructive oil spill.[14] These two accidents gave the conservationists a compelling argument that CRAMRA would not stop such accidents from devastating the Antarctic environment.

→ 'What took them so long': penguins after the Antarctic oil spill in 1989, by Australian cartoonist Geoff Pryor.

The treaty nations returned to the negotiating table and, over the next three years, developed the Protocol on Environmental Protection to the Antarctic Treaty (the Madrid Protocol) signed in 1991 by 26 treaty parties. The new protocol formally recognised Antarctica as a natural reserve devoted to peace and science, and established guidelines for conserving Antarctic flora and fauna, managing and disposing of waste, and preventing marine pollution. Significantly, it included an indefinite ban on all commercial mining in Antarctica for at least 50 years.[15] For Australia, the protocol was a significant diplomatic achievement and illustrated the nation's capacity to play a decisive leadership role on the world stage.[16]

The last wilderness

The disposal of unwanted materials in a perennially frozen continent had posed a particular challenge for Antarctic stations. While based on the continent, Greenpeace volunteers had undertaken more than 40 inspections of station environments, and what they found was shocking. At one station, they observed 'a pile of truck skeletons, wheels, oil drums and a pipe discharging brightly coloured liquid straight into the water'.[17] Since Mawson's Australasian Antarctic Expedition (AAE) in 1911, most of the waste produced on the Australian stations had either been burned, buried or dumped on sea ice during winter—a practice known as 'sea-icing', in which the rubbish, frozen until the summer melt, would eventually sink into the sea.[18] The problem was even more apparent on land, where disused building materials, packaging, machinery and fuel gradually accumulated around the stations. When ANARE moved to the first Casey station in 1969, the area around the old base at Wilkes was littered with some 3,000 rusting fuel drums containing oil and other chemicals. Richard Penney recalled how, when Australians first occupied Wilkes during the IGY, they had given little thought to the accumulating rubbish: 'We then felt we were still at the exploring stage so we didn't worry too much about the junk we left behind'. By the 1980s, however, there was a much greater awareness: 'And now we're beginning to see that this junk is still there; it doesn't go away, and the whole world is becoming more conscious of environmental issues'.[19]

→ Greenpeace made several voyages to Antarctica between 1986 and 1992 as part of a campaign to ban mining and protect the Antarctic continent. More recently, Greenpeace has called for the creation of an Antarctic Ocean sanctuary.

↑ Images such as this taken in 1987 drew international attention to the problems of pollution and rubbish around Antarctic research stations. The 1991 Madrid Protocol introduced strict regulations regarding waste management and the clean-up of hazardous materials to protect the Antarctic environment.

Even the new station at Casey had its critics. Stephen Murray-Smith visited in 1985 during its construction, and was alarmed:

Look ashore at man's inhumanity to Antarctica. Look at the tawdry reach-me-down, straggling, stepped buildings of our present base here, and a few hundred yards away up the rise the products of the plastic civilisation which is now replacing the tin civilisation ... Yes, we've gone a long way towards buggering up one continent. There's no reason why we shouldn't start on another.[20]

Casey is located near a rare ice-free oasis supporting an expanse of moss beds, lichens and algae, known colloquially as the 'Daintree of Antarctica.'[21] Ron Lewis-Smith, a Scottish botanist and expert on Antarctic mosses, who was on the voyage with Murray-Smith as an observer for the Antarctic Treaty, was unimpressed by the rubbish around the station. He said so in his report and, within the year, rubbish tips were closed at all Australian stations and the Antarctic Division began a concerted program of cleaning up and shipping all waste back to Australia.[22] Joan Russell, the station leader at Casey in 1990, organised clean-up days around the station, revealing just how quickly attitudes towards the fragile polar environment had changed:

I nearly drove everybody mad with the weekly emu bobs. We would take bags out, sector the station up and just comb it. You know, like emu bobs in the school yard, pick up every lolly paper. Well, we picked up every single piece of evidence of human habitation that we could find. It needs to go week after week after week after week.[23]

← Scientists are monitoring the health of moss beds near Casey station to understand the impact of climate change in East Antarctica.

↓ Sleeping tents at Law Dome with approaching storm, October 2004.

Reading the ice

In 2008, Australian physicist and glaciologist Dr Tas van Ommen found himself sheltering on Law Dome as a 120 kilometre per hour blizzard raged outside. His shelter was a tent mounted on a heavy wooden floor and fitted with aluminium sledges tethered to the ice. With each ferocious gust, the anchor ropes shook violently, threatening to send the whole edifice sliding and tumbling across the ice cap. Unable to sleep, van Ommen decided to make a dash for the mess tent:

> I grabbed a sleeping bag and literally felt like an explorer of old. I could hardly see where I was going ... into the main mess tent where we all camped for the next four or five days because it was just too hostile outside.[24]

He tells the story with relish, adding that such extreme conditions contributed to his enjoyment of being out in the field: 'I always think of myself as a semi-Antarctic being. It gets in your blood, I think'.

Douglas Mawson, for all his travails on the ice sheet, might well have agreed. Experiencing an Antarctic blizzard may not be for the faint-hearted, but the ice sheet has always had a way of infiltrating the soul. Standing on the crest of a ridge during one of his sledging journeys from Main Base at Commonwealth Bay during the AAE, Mawson felt an uncanny sense of timelessness:

> *Tramping over the plateau, where reigns the desolation of the outer worlds, in solitude at once ominous and weird, one is free to roam in imagination through the wide realm of human experience to the bounds of the great Beyond. One is in the midst of infinities—the infinity of the dazzling white plateau, the infinity of the dome above, the infinity of the time past since these things had birth, and the infinity of the time to come before they shall have fulfilled the Purpose for which they were created.*[25]

Understanding the purpose for which the 'dazzling white plateau' was created would become a central preoccupation of Antarctic science. Observers in Australia had been aware since the early twentieth century that the Antarctic region exerted a powerful influence on Australia's—and the world's—climate systems, although exactly how had remained a mystery.[26] By the 1980s, as international debates about the implications of rising levels of carbon dioxide (CO_2) intensified, Antarctic glaciologists began to shift the focus of their deep-field research from the physical characteristics of the ice sheet to exploring the past environmental conditions that lay within its depths. They knew that significant changes in climate patterns had profoundly affected past human societies. The Little Ice Age, which began in the early fourteenth century and lasted for some 500 years, was a particularly notable climatic event in human history, disrupting fisheries and agriculture and leading to widespread famine and dislocation of peoples across northern Europe.[27] The question was what the Antarctic ice sheet might reveal about the history of such climatic events. The answers lay in drilling deep into the ice sheet to retrieve long cylindrical cores from the ice formed thousands of years ago. Australia, as it happened, was one of the first nations in Antarctica to drill a long ice core from surface to bedrock, and the results were groundbreaking. The 1,200 metre core retrieved from Law Dome in the early 1990s contained a climate record stretching back more than 80,000 years.

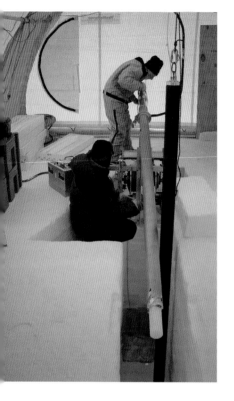

← Ice core drilling at Law Dome, 2004.

Law Dome, first photographed in 1946–1947 during Operation Highjump and initially called Wilkes ice cap, was an ideal place to undertake ice core research. Isolated from the rest of the plateau by the ocean to the north and two glaciers to the south, it represents a miniature version of the entire ice cap. Its location, near the coast at the edge of the East Antarctic Ice Sheet, means that it receives higher snowfalls than the interior. As a result, Law Dome contains an unusually detailed record of natural climate variability over the past 1,000 years, enabling climate scientists to estimate variations in Australia's rainfall before the introduction of rain gauges.[28] One of van Ommen's significant insights came when he compared the layers of ice with Australian meteorological records since 1900. The comparison revealed that an unusually snowy period in Antarctica coincided with prolonged drought conditions in southwestern Australia, and that such conditions were likely associated with climate change. For the first time, changes in snowfall recorded in the East Antarctic ice core over several centuries could be linked to rainfall patterns in Western Australia.[29]

For van Ommen, it was a moment to treasure as he stood on Law Dome, contemplating the archive of climate history beneath his feet:

> If you look back at Mawson and Edgeworth David and others, they stated as part of their mission that Antarctica is important for the climate of the Southern Hemisphere. But to actually find such a direct connection, which we did in the finish, was really potent ... To this day it's one of those kind of serendipity things. Normally the climate is a messy thing—it's got so many factors, especially in remote locations. You don't expect to find links as direct as that, but there it was ... watching the snow coming down around me and think that this is connecting

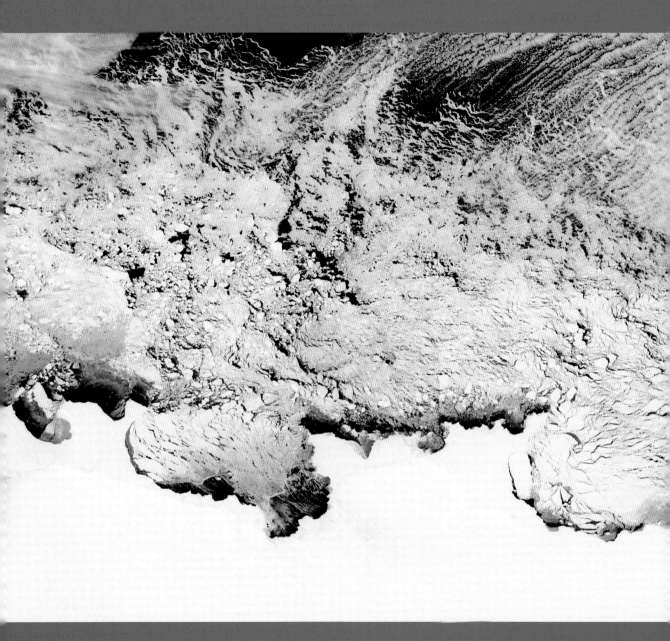

↑ A NASA satellite image of the Law Dome and the Knox, Budd and Sabrina coasts, taken as part of Operation Ice Bridge, the largest airborne survey ever undertaken to monitor the behaviour of Arctic and Antarctic ice sheets, ice shelves and sea ice.

me back to Australia and lower latitudes, you can put the whole thing together in
your head and it feels ... very connected.[30]

Van Ommen would spend a decade analysing the ice core in a Hobart laboratory before
setting foot on the ice, and it led to a career as an international expert in Antarctica's
paleoclimate and an abiding passion for Antarctica.

Ice cores contain bubbles of air, trapped as the snow falls and sealed off as it compresses,
forming 'little time capsules of past atmosphere'. The deep ice itself is so pure that it allows
scientists to measure tiny traces of dust from distant lands, carbon particles from forest
fires and dissolved impurities such as volcanic acids and sea salts, all present in Earth's
atmosphere at the time the snow fell. They can also determine snow temperatures and
levels of CO_2 present in the atmosphere, as well as information about seasonal cycles and
changes to annual snowfall. From the second half of the twentieth century, glaciologists
were learning to read Antarctica's frozen archive of past climatic conditions, tracking
changes over time by counting the layers off like tree rings. Plotted onto a simple graph,
it was possible to visualise Antarctica's ancient heartbeat measured in the regular rise
and fall of temperatures and CO_2 levels as the planet experienced intermittent ice ages
and interglacial warm periods. It also showed an alarming increase in Earth's CO_2 levels

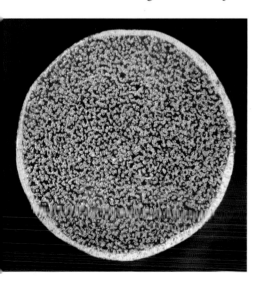

since the mid-twentieth century. As van Ommen noted in
2021, it was currently about 415 [CO_2 parts per million],
'so we are way above anything that's been seen in the
entire ice core and increasing by nearly three ppm per
year ... it's pretty potent stuff'.[31] Ice cores, extracted from
the most isolated and extreme environment on Earth,
have opened a window into deep time. The Antarctic ice
sheet is no timeless desert; it has its own rich stories to
tell, about how the world's climate has changed and how
it might change in the future.[32]

← Cross-section slice of an ice core showing trapped
bubbles of air from the atmosphere in the past.

↑ Glaciologist student Shavawn Donoghue sampling an ice core taken from Brown Glacier, Heard Island, in 2004.

→ A section of Antarctic ice core containing a layer of volcanic ash. The core is being analysed by the University of Washington for similarities with ice cores retrieved from Greenland.

It was only a matter of time before a decision was made to drill even further back into the past. In 2004, the European Project for Ice Coring in Antarctica (EPICA) completed drilling the longest continuous ice record on Dome C at Concordia station. The 800,000-year-old ice core showed how ice ages had recurred roughly every 100,000 years.[33] However, the much older ocean sediments retrieved by drilling into the seabed indicated that ice ages had previously occurred at around 40,000-year intervals. It seems that the pace changed about a million years ago, when the interval between ice ages slowed from 40,000 years to 100,000 years.[34] In order to understand this phenomenon, an international team of glaciologists co-led by van Ommen is preparing (at the time of writing) to drill the oldest ice core in the world. The million-year ice core project involves drilling 2,800 metres into the ice sheet in search of information about Earth's sensitivity to CO_2—information that no other place on Earth can provide. By examining levels of CO_2 and other gases contained in the deepest, oldest ice in Antarctica, the team seeks to solve the puzzle of what controls the pacing of the ice ages.

Finding the ideal site for the oldest ice in the world was no easy task. It needed to be, in Goldilocks' terms, thick enough but not so thick that the geothermal heat from Earth's crust could have melted the oldest ice at the bottom. Australia's search for a suitable site began in 2008 with a project called the International Collaboration for Exploration of the Cryosphere through Aerogeophysical Profiling (ICECAP). Using Basler aircraft (rebuilt old DC3 airframes) with radar equipment mounted under the wings, these modern-day explorer–scientists were able to 'see' through the ice sheet to measure its thickness and map the topography of the underlying bedrock that had so fascinated Mawson.[35] This exploration was an international collaboration with the United States and France, and also provided guidance to other nations' efforts.

The team finally settled on a site known as 'little Dome C', about 1,100 kilometres inland from Casey station and around 40 kilometres from the site of the 800,000-year ice core. Here, they predict that the bedrock will be sufficiently deep to hold the archives of a million years of snowfall, yet high enough to reduce the effects of geothermal heat. As van Ommen explained:

↑ Dome C at Concordia station was the scene of ice core drilling in 2004 to retrieve an 800,000-year-old ice core. In 2021–2022, work began at 'little Dome C' to drill an ice core exceeding one million years old in the quest to solve the mystery of changes in the cycle of past ice ages.

> The ice age pacing puzzle shows that we don't fully understand the long-term consequences of CO_2 that we're putting into the atmosphere but, when we look back in time, we start to get an understanding of what's driven the climate in the past and that allows us to predict better what's going to happen in the long-term future.[36]

Nevertheless, with annual air temperatures at 'little Dome C' averaging −55°C, the quest to drill deep into the ice sheet will test the limits of modern polar science and technology, just as the quest to explore its vast surface tested the limits of the earliest Antarctic expeditions more than a century ago.[37]

The tipping point

Even as the search is underway to decipher the secrets of the oldest glacial ice in the world, polar scientists are monitoring, in real time, how the ice sheet itself is responding to the warming of the ocean waters that surround it. The ice sheet maps produced by ICECAP revealed alarming new insights about the stability of Antarctica's ice sheet and the potential for it to contribute to rising global sea levels. When van Ommen's team mapped the area inland from Casey station, they discovered that a deep basin known as the Aurora Subglacial Basin was much more extensive than previously thought and connected close to the coast. It is a marine-based ice sheet, which, like the majority of West Antarctica, makes it potentially vulnerable to melting from warming ocean waters.[38] But it is the floating ice shelves extending along much of Antarctica's coastline that are at the frontline of global warming. These ancient custodians of the ice sheet presented a formidable barrier to early Antarctic voyagers. Perhaps the best known of these is the Ross Ice Shelf, the world's largest body of floating ice, measuring 600 kilometres in length and up to 50 metres in height. James Clark Ross, after whom it is named, originally called it the 'Great Ice Barrier' or simply 'The Barrier', during his expedition to the South Pole in 1841. These ice barriers may seem impenetrable, but they are far from stable.

← Antarctica's icebergs, calved from glaciers or ice shelves, are renowned for their spectacular shapes and colours, but places such as 'Iceberg Alley' also hold important clues for scientists studying how glacial ice has responded to past changes in Earth's climate.

Ice shelves are floating extensions of the ice sheet, forming as snow falls on the polar plateau and compresses into glacial ice that pushes seaward, flowing to the coast. They retreat as the ice melts and icebergs calve from their edges. This ebb and flow of glacial ice is Antarctica's supreme balancing act, but warming ocean waters are increasing the speed at which ice shelves are collapsing, with implications for the stability of large areas of the ice sheet.[39] When the Larsen B Ice Shelf collapsed in 2002, the conspicuous event captured global attention, with speculation about how much it was contributing to rising sea levels. One study found that, over the previous two decades, the ice shelf had become weakened by the loss of sea ice and increased exposure to the stormy swells of the Southern Ocean. Ice shelves are already afloat, so their collapse does not directly add to rising sea levels, but the disintegration of Larsen B released a torrent of glacial ice that had been held in place for more than 11,000 years.[40]

Changes in the behaviour of Antarctic marine and bird life in response to a changing environment may be less dramatic than the spectacle of a collapsing ice shelf, but such changes are equally disturbing. During one aerial reconnaissance of emperor penguins in 2010, seabird ecologist Dr Barbara Wienecke was shocked to find a colony huddled on the top of a grounded iceberg. The surrounding sea ice that year had simply been too weak to support the heavy birds, but the iceberg was far from stable and the location exposed the chicks to crevasses and strong winds. 'It nearly ripped my heart out,' she recalled. Four years later, aerial photographs and satellite imagery revealed four separate colonies rearing their chicks on ice shelves rather than on sea ice. Whether this was a previously unobserved behaviour or an adaptation to a changing climate was unclear, but it served to highlight the catastrophic future facing such highly specialised polar species on a warming planet.[41]

In 2002, one of Australia's most respected Antarctic geologists and palaeontologists, Patrick Quilty, predicted that Antarctica's importance would increase over the next 25 years—not only for Australia, but for the planet as a whole—as human-induced climate change increasingly impacted this sensitive polar region.[42] Recent studies of melting glaciers in Antarctica have shown the accuracy of this prediction. A 2018 study led by the Institute of Marine and Antarctic Studies at the University of Tasmania found

↑ Scientists were monitoring this rift system known as the 'loose tooth' in the Amery Ice Shelf for nearly two decades before an iceberg the size of Sydney broke away in 2019.

that the melting glacial ice in Antarctica is slowing ocean currents and, in turn, further accelerating the melting of the ice sheet.[43] For most Australians, however, warming ocean waters and collapsing ice shelves are far removed from daily life. As Tom Griffiths has written, Antarctica lies so far to the south that it seems to 'fall off the map', at least until a vortex of chilling polar wind carries the 'breath of Antarctica' to Australia's southern shores.[44] Nevertheless, as the stories in this book show, Australia and Antarctica are intimately linked by weather, ocean, geology, fauna and flora, and human history. The connections between these Gondwanan cousins run deep, and the future of this remote, forbidding continent and its circumpolar ocean will continue to be profoundly important for all Australians.

Acknowledgements

My thanks go to the many Australian collecting institutions, including archives, libraries, museums and galleries, for their generous assistance in providing access to their Antarctic and Southern Ocean collections. These include the National Library of Australia, the Australian Antarctic Division, the National Film and Sound Archive of Australia, the National Archives of Australia and the National Museum of Australia, as well as state libraries and museums of South Australia, Victoria, New South Wales and Tasmania. I am also grateful to the ANARE Club for making their archive available online. I would like to thank the many scholars and writers whose work has inspired and informed this book. I am especially grateful to Professor Tom Griffiths, whose expertise on Australia's Antarctic history helped me to navigate the complex environmental, political and cultural dimensions of the southern polar region, and who generously read the whole manuscript and provided such insightful and encouraging feedback. Professor Elizabeth Leane's extensive exploration of Antarctic culture and literature provided valuable insights, and Dr Tim Bowden's monumental ANARE Jubilee history contained a wealth of information and guidance to sources about the first 50 years of Australia's scientific program in the Antarctic region.

My ideas for this book have also been shaped by conversations with many people who, over the course of several years, have generously shared their memories, stories and perceptions of Antarctica and the Southern Ocean with me. I would especially like to acknowledge and thank Denise Allen, Dr Narissa Bax, Dr Jaimie Cleeland, Keith Gooley, Dr Tas van Ommen and Dr Barbara Wienecke for allowing me to interview them about their professional and personal experiences in the polar region and for generously sharing their stories and photographs. I would also like to thank those who kindly gave permission for me to quote them or to use extracts from published or unpublished works. In particular, I am grateful to Alun Thomas and Andrew McEwin, the Trustees of the Estate of the Late Sir Douglas Mawson, as well as the South Australian Museum for their generous permission to quote from Sir Douglas Mawson's writings. A wide range of scholars, scientists, artists and writers

have also contributed important ideas and advice, and I thank in particular James Bradley, Dr Andrew Constable, Dr Bernadette Hince, Dr Michael Pearson, Dr Lisa Roberts, Dr Elizabeth Truswell, and colleagues in the Scientific Committee on Antarctic Research (SCAR) Standing Committee on the Humanities and Social Sciences research community.

I am grateful to the Australian National University for supporting my continued association with the School of History as a visiting academic and member of the Centre for Environmental History. Professor Frank Bongiorno, Professor Carolyn Strange and Associate Professor Ruth Morgan have generously provided collegiate support and encouragement while enabling me to engage with colleagues and share my research with wider audiences.

The National Library of Australia's publishing imprint, NLA Publishing, commissioned this book and its team provided invaluable guidance and encouragement throughout the publishing process. Everyone at the Library has been unfailingly professional, enthusiastic and kind throughout an unusually challenging time and rapidly changing work environment. My thanks go especially to Susan Hall, Bobby Graham and Lauren Smith, to the project manager Amelia Hartney and to the Library staff members who assisted with this project, including Jemma Posch, Jessica Coates, Bronwyn Coupe and Kathryn Favelle. Staff of the Petherick Reading Room and collection specialists provided invaluable research assistance, especially in relation to the Library's oral history, manuscripts, map and photographic collections. The editorial skills of Amelia Hartney and Melody Lord and the design skills of Stan Lamond played a crucial role in shaping the final product.

My family, as always, has been steadfast in providing love and support throughout this project. I am grateful to Alan and Lucy for their care and encouragement, particularly as I began this project while recovering from surgery. I also thank them for being an engaged and thoughtful audience as I developed my ideas for the book. Finally, my thanks to friends, both old and new, who have contributed immeasurably by helping me to feel nurtured and grounded, especially as I moved house and states during the process of completing this book.

↓ Emperor penguins, Gould Bay, Antarctica.

Timeline

85–30 million years ago	The continental plate carrying Australia begins moving north. Antarctica becomes isolated by the Southern Ocean and the Antarctic Circumpolar Current begins circulating around the Southern Hemisphere. Australia's and Antarctica's modern environments evolve.
65,000 years ago	Earliest accepted date for human occupation of Australia, although recent research suggests even earlier occupation.
1772–1775	Captain James Cook circumnavigates the high southern latitudes and crosses the Antarctic Circle without sighting the Antarctic continent.
1819–1821	Russian naval officer Captain (later Admiral) Fabian Gottlieb von Bellingshausen circumnavigates the Antarctic continent and claims to have sighted it on 27 January 1820.
1820–1821	American sealer Captain John Davis claims to have made the first landing on the Antarctic continent.
1872–1876	*Challenger* expedition reaches Southern Ocean and crosses Antarctic Circle.
1886	Royal Society of Victoria establishes the Australian Antarctic Exploration Committee to investigate the establishment of research stations in Antarctica with the use of steam-powered vessels.
1894–1895	Norwegian whaler Henrik Johan Bull of Melbourne recruits Queensland-based teacher and scientist Carsten Egeberg Borchgrevink to voyage south on the *Antarctic*. He claims to have made the first landing on the continent (at Cape Adare).
1898–1900	Borchgrevink recruits Tasmanian physicist Louis Bernacchi on the *Southern Cross* expedition to Cape Adare. Bernacchi is the first Australian to land on the continent and the expeditioners are the first humans to winter there.
1901–1904	Bernacchi returns to Antarctica as physicist with Robert Falcon Scott's *Discovery* expedition.
1907–1909	Australians play a major role in Ernest Shackleton's *Nimrod* expedition to Cape Royds, including chief scientific officer Professor (later Sir) T.W. Edgeworth David, meteorological observers Leo Cotton and Bertram Armytage, geologist Douglas Mawson and master of the vessel Captain John King Davis. David and Mawson reach the south magnetic pole.
1910–1913	Australian geologists Thomas Griffith Taylor and Frank Debenham are members of Robert Falcon Scott's ill-fated *Terra Nova* expedition.
1911–1914	Mawson leads the Australasian Antarctic Expedition (AAE) to Antarctica, the first Australian-led expedition, aboard the *Aurora* with Captain Davis as his deputy. The AAE undertakes extensive exploration and scientific work and (in 1913) establishes the first radio contact between Antarctica and Australia.
1914–1917	Australian photographer Frank Hurley accompanies Ernest Shackleton's *Endurance* expedition, and his images of the ill-fated expedition earn him international recognition.
1915	Mawson publishes his account of the AAE in *The Home of the Blizzard*.

1929–1931   Mawson leads the British, Australian, New Zealand Antarctic Research Expeditions (BANZARE) to Heard Island, the Kerguelen Islands and the Antarctic coast aboard Scott's old ship *Discovery* under the command of Captain Davis (1929–1930) and Captain Mackenzie (1930–1931). Apart from Mawson and Davis, the Australians include Alf Howard (hydrologist), Stuart Campbell and Eric Douglas (aviators), Morton Moyes (cartographer) and Frank Hurley (photographer and cinematographer).

1930   Mawson declares a portion of the Antarctic continent as British sovereign territory.

1933   Britain transfers its Antarctic territorial claim to Australia under the *Australian Antarctic Territory Acceptance Act 1933* (effective 1936). Tasmanian government revokes sealing licenses on Macquarie Island and declares it to be a wildlife sanctuary.

1947   The Australian government appoints an Executive Planning Committee to oversee a program of exploration and research in Antarctica—the Australian National Antarctic Research Expedition (ANARE)—building on Mawson's earlier work in Antarctica and consolidating Australia's interest in the Australian Antarctic Territory (AAT).

1947   Australia claims Heard Island as part of the AAT and establishes a station at Atlas Cove to be manned by 14 winterers (Heard Island station is closed in 1955).

1948   ANARE establishes a research station on Macquarie Island. ABC radio program *Calling Antarctica* begins broadcasting.

1949–1966   Dr Phillip Law serves as the first director of the Antarctic Division and leader of ANARE. Known as 'Mr Antarctica', Law is influential in shaping Australia's early scientific program and station culture in Antarctica.

1954   *Kista Dan* sails from Melbourne on 4 January on a mission to establish a permanent base in the AAT. Law selects a site on Horseshoe Harbour, and Mawson research station is commissioned on 13 February with ten winterers to spend the first year constructing buildings, undertaking meteorological observations and exploring coastal and inland regions. Husky teams are introduced to Mawson station.

1956   The Royal Australian Air Force (RAAF) conducts the first flights to photograph the Antarctic coast. Coal is discovered in the Prince Charles Mountains.

1957   Davis research station is established on 13 January to support Australian scientists undertaking research for the International Geophysical Year (IGY) in 1957–1958. The United States establishes Wilkes station.

1957–1958   The IGY officially commences on 1 July 1957, with 12 nations (including Australia) engaged in Antarctic expeditions.

1958   ANARE, aboard *Thala Dan*, explores almost the entire AAT coast and Adélie Land, and establishes the first continuous automatic weather stations in Antarctica on Lewis Island and Chick Island. ANARE employs helicopters for the first time.

1959   Twelve nations (including Australia), whose scientists were active in Antarctica during the IGY, sign the Antarctic Treaty.

1959   The US transfers Wilkes station to Australia.

1959   The first four women to be permitted to visit an ANARE station—Australians Dr Isobel Bennett, Hope McPherson and Susan Ingham, along with British biologist Mary Gilham—arrive at Macquarie Island on 21 December to undertake research as part of the summer science program.

1961	Nel Law, wife of Phillip Law, arrives at Mawson as a guest aboard *Magga Dan*. She is the first Australian woman to visit the continent and the first woman to set foot on an Australian continental station.
1962	Six ANARE expeditioners begin a 17-week 3,000 kilometre traverse from Wilkes to Russia's Vostok station and back.
1965–1969	With Wilkes buried in snow, construction begins on a replacement station known as Repstat. Davis station closes for four years to free up resources. The new station is officially opened on 19 February 1969 and called Casey research station.
1968	Four ANARE expeditioners winter on the Amery Ice Shelf to measure the speed and flow of the ice flowing from the Lambert Glacier, and ANARE begins an extensive field survey of the Prince Charles Mountains.
1971	ANARE makes the first recorded landing on the McDonald Islands near Heard Island on 27 January (the first survey is undertaken in 1979–1980).
1972	Seventeen nations, including Australia, ratify the Convention for the Conservation of Antarctic Seals (enters into force in 1978).
1972	ANARE opens an underground cosmic ray observatory at Mawson and drilling commences in the ice sheet on Law Dome.
1975–1976	The first three Australian women to visit a continental station under the ANARE program arrive at Casey. Dr Zoë Gardner becomes the first woman to winter with ANARE, serving as medical officer on Macquarie Island.
1977	Qantas conducts the first of a series of one-day tourist charter flights over Antarctica. A second series is not conducted until 1994.
1978	RAAF begins flights between Christchurch in New Zealand and the American McMurdo station as part of a cooperative air transport system in Antarctica. Scientists and summer personnel arrive on a United States Hercules at Lanyon Junction for summer at Casey.
1980	Australia signs the Convention on the Conservation of Antarctic Marine Living Resources (CCAMLR) after terms are agreed at a conference of Antarctic Treaty nations in Canberra (in force from 1982).
1980–1981	ANARE's first dedicated marine biology research is conducted aboard *Nella Dan*, which has been modified for the purpose. The first marine geophysical survey is conducted in Prydz Bay in January 1982.
1981	Dr Louise Holliday is the first woman to winter on an Australian continental station, serving as medical officer at Davis.
1984	*Icebird* leaves Cape Town for its first ANARE voyage, enabling more expeditioners to be transported to Australia's Antarctic bases.
1984–1985	Dr Patrick Quilty discovers fossil dolphin bones at Marine Plain near Davis.
1986	ANARE establishes the first of a series of summer bases to support its deep field activities. The base in the Bunger Hills is named Edgeworth David after the Australian geologist who accompanied Shackleton's *Nimrod* expedition and Mawson's AAE.

1987   *Nella Dan* runs aground at Macquarie Island and is scuttled in deep water.

1987   Greenpeace establishes a base at Cape Evans as part of an environmental campaign to establish an Antarctic World Park.

1987–1988   ANARESAT satellite communications terminals are installed at Australia's Antarctic stations.

1988   Diana Patterson becomes the first female station leader of an Australian Antarctic station, serving as OIC of Mawson.

1988   Australians view the first live television broadcast from the AAT (at Davis station) as part of Australia's Bicentennial celebrations, and 'new Casey' is opened as part of a major rebuilding program across Australia's three continental stations.

1988   As Antarctic Treaty nations prepare to sign the Convention on the Regulation of Antarctic Mineral Resource Activities (CRAMRA), Australians are divided over its merits.

1989   *Aurora Australis* is launched at Newcastle and is chartered by ANARE as a resupply and research vessel.

1991   Australia and France join forces to urge Antarctic Treaty nations to develop a new environmental regime that bans mining in Antarctica. Following diplomatic efforts and two devastating oil spills in Antarctica and Alaska, Treaty nations agree to the Protocol on Environmental Protection to the Antarctic Treaty (Madrid Protocol) designating Antarctica a 'natural reserve, dedicated to peace and science' (Article 2).

1992–1993   ANARE glaciologists retrieve a 1,200 metre ice core from Law Dome revealing past climatic conditions.

1993   The last of the huskies on Australia's stations are removed from Mawson station as required under the Madrid Protocol.

1994   Australia adopts a new science program that determines the future direction of its Antarctic research. It moves from discipline-based to multidisciplinary science to support Australia's policy commitments in Antarctica.

1995   A new Antarctic Climate and Ecosystems Cooperative Research Centre significantly increases Australia's marine research activities, and a new Antarctic Data Centre is established to coordinate data and make it more accessible to international researchers. ANARE undertakes the first voyage into sea ice in winter.

1997   ANARE celebrates its 50th anniversary. Macquarie Island, and the Territory of Heard Island and McDonald Islands are inscribed on the World Heritage List.

2007–2009   Australia leads field research in Antarctic biology, sea ice and climate science during the Fourth International Polar Year.

2010   Australian glaciologists retrieve ice cores from Law Dome that show the link between the Australian and Antarctic climates. Australia adopts a new science strategic plan for the next ten years.

2021   Australia's new purpose-built icebreaker, *Nuyina*, arrives in Hobart, marking the beginning of a new era of scientific research in Antarctica and the Southern Ocean.

Notes

Preface

1   Commonwealth of Australia, *Antarctic Treaty. Report of the First Consultative Meeting*, 1961, p.30.
2   Tom Griffiths, 'Introduction: Listening to Antarctica', in Bernadette Hince et al. (eds), *Antarctica: Music, Sounds and Cultural Connections*, Canberra: ANU Press, 2015, p.7.

Chapter 1: Continent

1   Douglas Mawson, *The Home of the Blizzard*, vol.1, London: William Heinemann, 1915, p.157.
2   South America is the closest continent to Antarctica: the tip of the Antarctic Peninsula lies 1,238 kilometres south of Ushuaia in Argentina. For a table of distances between Australian cities and Antarctic stations, see Australian Antarctic Division, 'Distances', antarctica.gov.au/about-antarctica/geography-and-geology/geography/distances/.
3   This record was broken in 2018 when scientists used satellite technology to record –97.8°C on the East Antarctic Plateau, clost to the theoretically coldest temperature that Earth can reach. Caleb A. Scharf, 'The Coldest Place on Earth', *Scientific American*, 29 January 2019, blogs.scientificamerican.com/life-unbounded/the-coldest-place-on-earth/; Australian Antarctic Division, 'Dome Argus', updated 16 February 2018, antarctica.gov.au/antarctic-operations/stations/other-locations/dome-a/.
4   Australian News and Information Bureau, Antarctic Division, *Australians in the Antarctic*. Produced for the Antarctic Division of the Department of External Affairs by the Australian News and Information Bureau, Melbourne, 1961.
5   Andrew Klekociuk and Barbara Wienecke, Commonwealth of Australia, *Australia State of the Environment 2016: Antarctic Environment*, Independent report to the Australian Government Minister for the Environment and Energy, Canberra: Australian Government, 2017, pp.3–5.
6   Australian Antarctic Division, 'Antarctica divided into distinct biogeographic regions', antarctica.gov.au/news/2012/antarctica-divided-into-distinct-biogeographic-regions/.
7   Vinson Massif, *Encyclopedia Britannica*, britannica.com/place/Vinson-Massif.
8   Australian Antarctic Division, 'Antarctic Geology', antarctica.gov.au/about-antarctica/geography-and-geology/geology/.
9   'The Beginning Island: A Creation Story from the First Time, When Tasmania Was Born' (retold with permission by Pauline E. McLeod) in Helen F. McKay (ed.), *Gadi Mirrabooka: Australian Aboriginal Tales from the Dreaming*, Eaglewood: Libraries Unlimited, 2001, pp.39–40; M. Gantevoort et al., 'Reconstructing the Star Knowledge of Aboriginal Tasmanians', *Journal of Astronomical History and Heritage*, vol.193, no.3, 2016, pp.327–347.
10 Tom Griffiths, 'The Breath of Antarctica', in *Tasmanian Historical Studies*, vol.11, 2006, pp.4–14; Tom Griffiths, *Slicing the Silence: Voyaging to Antarctica*, Sydney: NewSouth, 2007, p.4.
11 R.A. Swan, *Australia in the Antarctic: Interest, Activity and Endeavour*, Carlton: Melbourne University Press, 1961, p.33.
12 John Zillman, *A Hundred Years of Science and Service: Australian Meteorology Through the Twentieth Century*, Australian Bureau of Statistics, 1301.0 – Yearbook Australia, 2001, pp.4–5. The first weather map was published in a newspaper in 1877.
13 Henry Chamberlain Russell, 'Moving Anticyclones in the Southern Hemisphere', in Ralph Abercromby (ed.), *Three Essays on Australian Weather*, Sydney: F.W. White, 1896, pp.1–15; Griffith Taylor, 'Climatic Relations between Antarctica and Australia', in W.L.G. Joerg (ed.), *Problems of Polar Research*, New York: American Geographical Society Special Publication No. 7, 1928, pp.286–300; Tom Griffiths, 'The Roaring Forties', in Tim Sherratt et al. (eds), *A Change in the Weather: Climate and Culture in Australia*, Canberra: National Museum of Australia Press, 2005, p.160.
14 Clement Wragge, cited in Gavin Souter, *Acts of Parliament: A Narrative History of the Senate and House of Representatives*, Carlton: Melbourne University Press, 1988, p.8.
15 J.W. Gregory, *The Climate of Australasia in Reference to Its Control by the Southern Ocean*, Melbourne: Whitcombe and Tombs Ltd, 1904, p.30.
16 J.M. Powell, 'Taylor, Thomas Griffith (1880–1963)', *Australian Dictionary of Biography*, vol.12, Carlton: Melbourne University Press, 1990.

17 Griffith Taylor became a renowned geographer whose views on the environmental limitations on Australia's development made him a controversial figure in his day. See Carolyn Strange and Alison Bashford, *Griffith Taylor: Visionary, Environmentalist, Explorer,* Toronto: University of Toronto Press and Canberra: National Library of Australia, 2008; Powell, 'Taylor, Thomas Griffith (1880–1963)', *ADB*; Papers of Thomas Griffith Taylor (1880–1963), National Library of Australia, MS 1003.

18 Apsley Cherry-Garrard, *The Worst Journey in the World,* London: Vintage, 2010 (1922), p.318.

19 Australian Antarctic Division, 'Aurora', antarctica. gov.au/about-antarctica/environment/atmosphere/ aurora.

20 Duane W. Hamacher, 'Aurorae in Australian Aboriginal Traditions', *Journal of Astronomical History and Heritage,* vol.17, no.1, 2014, pp.207–17; Duane W. Hamacher, 'Fire in the Sky: The Southern Lights in Indigenous Oral Traditions', *The Conversation,* 2 April 2015, theconversation.com/ fire-in-the-sky-the-southern-lights-in-indigenous-oral-traditions-39113. See also Elizabeth Leane, *South Pole: Nature and Culture,* London: Reaktion Books Ltd, 2016, pp.32–33.

21 *The Hobart Town Courier,* 20 November 1835, p.2.

22 'The Aurora Australis', *The Argus* (Melbourne, Victoria), 3 September 1859, p.4.

Chapter 2: Ice

1 Mawson, *The Home of the Blizzard,* vol.1, pp.52–53.

2 Claudius Ptolemy, *Geographia,* trans. Giacomo d'Angelo da Scarperia, Bologna: Dominicus de Lapis, 1477 (c. 150 CE). The Greek geographer Marinus of Tyre coined the term 'Antarctic' in the second century CE to describe an imaginary southern polar area, the opposite of the Arctic region to the north.

3 A world map produced by Abraham Ortelius showed a landmass extending from the South Pole to the Tropic of Capricorn, reflecting European ideas about the geographical nature of the world in the sixteenth century. The map formed part of a series published in *Theatrum Orbis Terrarum* (Theatre of the World), 1570, Library of Congress, Control no.98687183.

4 'Secret Instructions for Lieutenant James Cook Appointed to Command His Majesty's Bark the *Endeavour*', 30 July 1768, National Library of Australia, MS 2.

5 James Cook, *A Voyage towards the South Pole, and round the World, Performed in His Majesty's Ships the Resolution and Adventure, in the Years 1772, 1773, 1774, and 1775,* vol.1, book 1, June 1772, London: W. Strahan & T. Cadell, 1777, Chapter II.

6 Cook, *A Voyage towards the South Pole,* Chapter II.

7 Joy McCann, *Wild Sea: A History of the Southern Ocean,* Sydney: NewSouth, 2018, pp.67–71.

8 The populations gradually recovered during the twentieth century, with an estimated 1.5 million fur seals recorded in 1993. George A. Knox, *Biology of the Southern Ocean,* 2nd edition, Boca Raton: CRC, 2007, p.216.

9 Michael Nash, *The Bay Whalers: Tasmania's Shore-based Whaling Industry,* Woden: Navarine, 2003, pp.127–160; Susan Lawrence and Mark Staniforth (eds), *The Archaeology of Whaling in Southern Australia and New Zealand,* Gundaroo: Brolga Press for Australasian Society for Historical Archaeology and Australian Institute for Maritime Archaeology, 1998.

10 Lorne K. Kriwoken and John W. Williamson, 'Hobart, Tasmania: Antarctic and Southern Ocean Connections', *Polar Record,* vol.29, no.169, 1993, pp.93–102.

11 Elizabeth Truswell, *A Memory of Ice: The Antarctic Voyage of the Glomar Challenger,* Acton: ANU Press, 2019, pp.5–7; G.E. Fogg, *A History of Antarctic Science,* Cambridge: Cambridge University Press, 1922, p.248.

12 Australian Antarctic Division, 'Antarctic Prehistory', antarctica.gov.au/about-antarctica/geography-and-geology/geology/antarctic-prehistory; 'Beardmore Glacier', *Encyclopedia Britannica,* updated 12 January 2007, britannica.com/place/Beardmore-Glacier; Truswell, *A Memory of Ice,* pp.8–14.

13 Tom Griffiths, 'The Breath of Antarctica', *Tasmanian Historical Studies,* vol.11, 2007, pp.6–7 [4–14]

14 *Sydney Gazette and New South Wales Advertiser,* 15 April 1820, p.3. The article cites 'Captain Billinghausen' and the ship 'Wostock'. I am indebted to Maria Kawaja's unpublished PhD thesis for identifying many useful primary sources dealing with Antarctica and the Australian colonies prior to Federation. See Marie Kawaja, 'The Politics and Diplomacy of the Australian Antarctic, 1901–1945', PhD thesis, Australian National University, 2010, Chapter 1.

15 Douglas Mawson, 'Presidential Address: The Unveiling of Antarctica', *Report of the Twenty-Second Meeting of the Australian and New Zealand Association for the Advancement of Science,*

Melbourne, January 1935, p.12; *Sydney Gazette and New South Wales Advertiser*, 2 June 1831, p.2.

16 *Commercial Journal and Advertiser*, 18 April 1840, p.2; *Hobart Town Advertiser*, 28 February 1840, p.3.

17 *The Courier* (Hobart, Tas), 4 May 1841, p.2. The play, 'The South Polar Expedition', was performed at the Royal Victoria Theatre on 3 May 1841.

18 *The Sydney Morning Herald*, 25 April 1891, p.8.

19 John Cawood, 'The Magnetic Crusade: Science and Politics in Early Victorian Britain', *Isis*, vol.70, no.4, 1979, pp.492–518.

20 *Antarctic Exploration: Baron Nordenskiöld's Proposal for Further Antarctic Exploration. With notes on the same by Alfred J Taylor*, Hobart, 1890, p.1.

21 Henrik Johan Bull, *The Cruise of the 'Antarctic' to the South Polar Regions*, London and New York: Edward Arnold, 1896, pp.44–45.

22 Bull, *The Cruise of the 'Antarctic'*, p.180. Other claims include the American sealing vessel *Cecilia* in 1821.

23 Clements R. Markham, 'The Need for an Antarctic Expedition', in *The Nineteenth Century: A Monthly Review*, March 1877–December 1900, London, vol.38, issue 224, October 1895, p.706 [706–712]; R.A. Swan, 'Borchgrevink, Carsten Egeberg (1864–1934)', *Australian Dictionary of Biography*, vol.7, Carlton: Melbourne University Press, 1979; 'Sixth International Geographical Congress, London, 1895', in *Journal of the American Geographical Society of New York*, vol.27, no.3, 1895, p.298.

24 Louis Charles Bernacchi, *To the South Polar Regions: Expedition of 1898–1900*, London: Hurst and Blackett, 1901, p.32.

25 Bernacchi's diary entry, 16 February 1899, cited in Janet Crawford [Bernacchi's granddaughter], *That First Antarctic Winter: The Story of the Southern Cross Expedition of 1898–1900 As Told in the Diaries of Louis Charles Bernacchi*, Christchurch: South Latitude Research Limited/Peter J Skellerup, 1998, p.81.

26 Crawford, *That First Antarctic Winter*, p.87.

27 Crawford, *That First Antarctic Winter*, p.109.

28 Bernacchi's diary entry, 18 June 1899, cited in Crawford, *That First Antarctic Winter*, p.128.

29 Bernacchi's diary entry, 25 December 1899, cited in Crawford, *That First Antarctic Winter*, p.171.

30 Borchgrevink's account of the expedition was published in 1901 as *First on the Antarctic Continent*. Bernacchi, *To the South Polar Regions*, pp.287–308; also see R.A. Swan, 'Bernacchi, Louis Charles (1876–1942)', *Australian Dictionary of Biography*, vol.7, Carlton: Melbourne University Press, 1979;

Crawford, *That First Antarctic Winter*.

31 Dana Bergstrom and Sharon A. Robinson, 'Casey: the Daintree of Antarctica', 2010, ro.uow.edu.au/scipapers/434; Australian Antarctic Division, 'Casey: the Daintree of Antarctica', antarctica.gov.au/about-antarctica/plants/casey-the-daintree-of-antarctica.

32 Melinda Waterman et al., 'Antarctica's "Moss Forests" Are Drying and Dying', *The Conversation*, 25 September 2018, theconversation.com/antarcticas-moss-forests-are-drying-and-dying-103751. Also see Truswell, *A Memory of Ice*, pp.206–208.

33 Marie Kawaja, 'Australia in Antarctica: Realising an Ambition', *The Polar Journal*, vol.3, no.1, 2013, p.35 [31–52].

34 Sir Ernest Henry Shackleton, *The Heart of the Antarctic: Being the Story of the British Antarctic Expedition 1907–1909, by E.H. Shackleton, C.V.O., with an introduction by Hugh Robert Mill, DSc, an account of the first journey to the south magnetic pole by Professor T.W. Edgeworth, F.R.S.*, London: W. Heinemann, 1909, p.31; David Branagan, *T.W. Edgeworth David: A Life*, Canberra: National Library of Australia, 2004, p.144. Among Shackleton's recruits were two of David's students—Douglas Mawson and Leo Cotton—as well as a British-born Australian seaman, Captain J.K. Davis, appointed as chief officer aboard the steam yacht. Davis subsequently served as master of the *Aurora* and Mawson's second-in-command during the AAE in 1911–1914, and later as commander of the Ross Sea Relief Expedition to rescue Shackleton's 'shore party' from McMurdo Sound in 1916–1917, and commander of the *Discovery* during the first voyage of Mawson's BANZARE (1929–1930). He received the King's Polar Medal for his services to polar exploration, and Australia's second continental station, established in 1957, was named in his honour.

35 Edgeworth David, cited in Gordon Greenwood and Charles Grimshaw (eds), *Documents on Australian International Affairs 1901–1918*, West Melbourne: Thomas Nelson (Australia) with the Australian Institute of International Affairs and the Royal Institute of International Affairs, London, 1977, pp.549–551.

36 Shackleton, *The Heart of the Antarctic*.

37 Edgeworth David, 'Narrative of the Magnetic Pole Journey', in Shackleton, *The Heart of the Antarctic*.

38 Edgeworth David, *The First Journey to the South Magnetic Pole: Part of the Story of the British*

Antarctic Expedition 1907–1909, Philadelphia: The Washington Square Press, 1909, p.129.

39 Edgeworth David, 'Extracts from the Narrative of Professor David', in Shackleton, *The Heart of the Antarctic*, p.310.

40 Edgeworth David, cited in F.J. Jacka, 'Mawson, Sir Douglas (1882–1958)', *Australian Dictionary of Biography*, vol.10, Calrton: Melbourne University Press, 1986. Their sledging record remained unbroken until the 1980s.

Chapter 3: Coast

1 Mawson, *The Home of the Blizzard*, vol.1, p.134.

2 Marie Kawaja, 'Australia in Antarctica: Realising an Ambition', *The Polar Journal*, vol.3, no.1, 2013, pp.37 and 49.

3 Mawson, *The Home of the Blizzard*, vol.1, p.xiv. Of the 36 scientific staff and crew of the *Aurora*, 20 were from Australia and the remainder from the United Kingdom, New Zealand and Switzerland.

4 John Robert Francis (Frank) Wild had established his reputation in Antarctica with Ernest Shackleton on the *Discovery* and *Nimrod* expeditions. After returning from the AAE, he served as his second-in-command on the ill-fated Imperial Trans-Antarctic expedition when he led the party that survived in upturned whaleboats for 105 days on Elephant Island after the loss of the *Endurance*.

5 John Béchervaise, 'Davis, John King (1884–1967)', *Australian Dictionary of Biography*, vol.8, Carlton: Melbourne University Press, 1981.

6 See Mawson, *The Home of the Blizzard*, vol.2, Appendix 1 for a full list of the expeditioners and crew.

7 Charles Francis Laseron, Diaries, 21 November 1911–24 February 1913, kept while a member of the Australasian Antarctic Expedition, 1911–1914, together with related papers, 1911–c. 12 October 1912, 13 December 1911. Mitchell Library, State Library of New South Wales, ML MSS 385.

8 Laseron, Diaries, 13 January 1912.

9 Charles Francis Laseron, *South with Mawson: Reminiscences of the Australasian Antarctic Expedition, 1911–1914*, London and Sydney. George G. Harrap and Company Ltd in association with Australasian Publishing Co Pty Ltd, 1947.

10 J.K. Davis, *High Latitude*, Parkville: Melbourne University Press, 1962, p.169.

11 Cecil T. Madigan, 'Tabulated and Reduced Records of the Cape Denison Station, Adélie Land', in Australasian Antarctic Expedition 1911–1914, *Scientific Reports*, series B, vol.IV, Meteorology, Sydney: Alfred J, Kent, Government Printer, 1929, p.17.

12 See Tom Griffith's evocative description of daily weather observations during the AAE in Griffiths, *Slicing the Silence*, pp.47-50.

13 Madigan, 'Tabulated and Reduced Records of the Cape Denison Station, Adélie Land', p.20.

14 Mawson, *The Home of the Blizzard*, vol.1, pp.115–116, 120.

15 Mawson, *The Home of the Blizzard*, vol.1, p.129.

16 Mawson, *The Home of the Blizzard*, vol.1, pp.119–120.

17 Laseron, Diaries, 4 February 1913.

18 Frank Wild, Papers, c. 1921–1937, vol.1, Memoirs, 1937?, Mitchell Library, State Library of New South Wales, ML MSS 2198, p.128.

19 Wild, Papers, c. 1921–1937, vol.1, Memoirs, 1937?, p.142.

20 Douglas Mawson, *AAE Scientific Reports A, vol.1, Narrative and Cartography*, Sydney: Government Printer, 1943.

21 Frank Wild, 'The Western Base—Winter and Spring', in Mawson, *The Home of the Blizzard*, vol.II, p.65.

22 Finnesko is a boot made of tanned reindeer skin with hair on the outside. Mawson, *The Home of the Blizzard*, vol.1, p.135.

23 Mawson, *The Home of the Blizzard*, vol.1, p.135.

24 Mawson, *The Home of the Blizzard*, vol.1, p.144.

25 Mawson, *The Home of the Blizzard*, vol.1, p.146.

26 Mawson, *The Home of the Blizzard*, vol.1, p.168.

27 Elizabeth Leane, *Antarctica in Fiction: Imaginative Narratives of the Far South*, Cambridge: Cambridge University Press, 2012, p.132.

28 Laseron, Diaries, 4 February 1913. Laseron participated in two sledging journeys between 8 November 1912 and 6 January 1913, and left Antarctica on 8 February 1913 for the return voyage to Australia.

29 Laseron, Diaries, 12 October 1912; Laseron, *South with Mawson*, p.110.

30 Leane, *Antarctica in Fiction*, pp.114–116; Isabella Beeton, *Mrs Beeton's Book of Household Management*, London: S.O. Beeton Publishing, 1861.

31 Charles Turnbull Harrisson, Diary, 23 August 1912. Mitchell Library, State Library of New South Wales, ML MSS 386. The Western Party comprised the leader, Frank Wild, George Harris Sarjeant Dovers (cartographer), Charles Turnbull Harrisson (biologist

and artist), Charles Archibald Hoadley (geologist), Sydney Evan Jones (medical officer), Alexander Lorimer Kennedy (magnetician), Morton Henry Moyes (meteorologist) and Andrew Dougal Watson (geologist).

32 Harrisson, Diary, 23 August 1912.

33 Harrisson, Diary, 1 September 1912.

34 R.L. Stevenson, *The Letters of Robert Louis Stevenson to His Family and Friends*. Selected and edited with notes and introduction by Sidney Colvin, London: Methuen and Co, 1899.

35 Harrisson, Diary, 8 September 1912. Harrisson's party returned to the Western Party hut on 15 September, then he was away again a week later on a second sledging journey.

36 Harrisson, Diary, 30 October 1912.

37 Morton Henry Moyes, Diary, 2 December 1911– 23 February 1913, p.154, Mitchell Library, State Library of New South Wales, ML MSS 388/Box 1/ Item 1. Moyes's diary includes a list of books he read during the expedition.

38 The air-tractor towed a train of four sledges, but it was abandoned after just 16 kilometres and later dragged back to Cape Denison, where it remains buried under snow.

39 Mawson, *The Home of the Blizzard*, vol.1, p.216.

40 Mawson's leader's brief to Frank Wild, written at Commonwealth Bay, 18 January 1912, cited in 'Home of the Blizzard', Australian Antarctic Division, updated 15 November 2011, mawsonshuts. antarctica.gov.au/western-party/new-lands.

41 Frank Hurley, Sledging diary, 10 November 1912–10 January 1913, kept while a member of the Australasian Antarctic Expedition, 1911–1914, together with an edited typescript transcript, Mitchell Library, State Library of New South Wales, ML MSS 389/1.

42 Francis Bickerton, cited in Mawson, *The Home of the Blizzard*, vol.2, p.9.

43 The 'Adélie Land meteorite', was the first of thousands of meteorites subsequently found on the Antarctic continent. Mawson donated it to the Australian Museum in 1924. See Bickerton, cited in Mawson, *The Home of the Blizzard*, vol.2, p.11.

44 Mawson, *The Home of the Blizzard*, vol.2, pp.3–4.

45 Mawson, *The Home of the Blizzard*, vol.1, pp.239–240.

46 Mawson, *The Home of the Blizzard*, vol.1, p.245. Mawson had calculated that the first sledge would probably suffer in the event of a snow-bridge collapsing, so the strongest dogs and the bulk of

their rations and equipment were on Ninnis's sledge at the rear.

47 Until his own death in 1958, Mawson never knew what had caused hair and skin loss and stomach pains, and speculation continues as to whether their bodies were reacting to toxic levels of vitamin A as a result of eating the dogs' livers.

48 Mawson, *The Home of the Blizzard*, vol.1, p.259.

49 Mawson, *The Home of the Blizzard*, vol.1, p.265.

50 Stillwell's diary entries for Wednesday 22 January 1913 and Monday 27 January 1913, cited in Bernadette Hince (ed.), *Still No Mawson: Frank Stillwell's Antarctic Diaries 1911–13*, Canberra: Australian Academy of Science, 2012, pp.216–217.

51 Carolyn Philpott and Elizabeth Leane, 'Making Music on the March: Sledging Songs of the "Heroic Age" of Antarctic Exploration', *Polar Record*, vol.52, no.6, 2016, p.698 [698–716].

52 Frank Hurley, 10 January 1913. Sledging diary, 10 November 1912–10 January 1913, from Frank Hurley diary collection, Mitchell Library, ML MSS 389/1–5; 'Southern Sledging Song', *Adelie Blizzard*, p.2.

53 Frank Hurley, *Argonauts of the South: Being a Narrative of Voyagings and Polar Seas and Adventures in the Antarctic with Sir Douglas Mawson and Sir Ernest Shackleton*, New York and London: G.P. Putnam's Sons, 1925, p.93.

54 Carolyn Philpott, 'Sledging Songs, Penguins and Melting Ice: How Antarctica Has Inspired Australian Composers', *The Conversation*, 29 August 2018, theconversation.com/sledging-songs-penguins-and- melting-ice-how-antarctica-has-inspired-australian- composers-101325.

55 Mawson, *The Home of the Blizzard*, vol.2, p.131.

56 Until recently, Jeffryes' achievements at Commonwealth Bay were overshadowed by Mawson's views on Jeffryes' mental illness, but his contribution to Australian Antarctic history has been restored in more sympathetic accounts. See for example: Elizabeth Leane et al., 'Beyond the Heroic Stereotype: Sidney Jeffryes and the Mythologising of Australian Antarctic History', *Australian Humanities Review*, vol.64, pp.1–23; and 'Sydney Jeffryes remembered', *Australian Antarctic Magazine*, Issue 35, December 2018, antarctica.gov.au/magazine/ issue-35-december-2018/in-brief/sidney-jeffryes- remembered.

57 Tom Griffiths, 'Role of Australians in Antarctica, 1913', for Michelle Hetherington (ed.), *Glorious Days: Australia 1913*, Canberra: National Museum of

Australia, 2013. Unpublished paper prepared for the NMA exhibition, supplied by the author.

58 Mawson, *The Home of the Blizzard*, vol.2, p.166.

59 'The Frozen South. The Australian [sic] Antarctic Expedition. Lecture by Sir Douglas Mawson', *The Advertiser* (Adelaide), 8 September 1914, p.10.

60 'Mawson's Expedition to the Ice-bound South', *The Observer* (Adelaide), 7 March 1914, p.28.

61 J.G. Hayes, *Antarctica: A Treatise on the Southern Continent*, London: Richards Press, 1928, p.22.

62 A. Grenfell Price, *The Winning of Australian Antarctica: Mawson's BANZARE Voyages, 1929–31, based on the Mawson papers*, published by Mawson Institute for Antarctic Research, University of Adelaide, Sydney: Angus and Robertson, 1962.

63 'Proclamation of Sovereignty Rights Over Antarctic Territory Made by Sir Douglas Mawson', 13 January 1930, National Archives of Australia, B1759, MAWSON B.

64 Australia is one of seven countries to have established territorial claims in Antarctica (the others include Argentina, Chile, France, New Zealand, Norway and the United Kingdom) although some countries do not recognise any territorial claims. The area known as the Australian Antarctic Territory is nearly 80 per cent of the size of the Australian continent, including all islands and territories south of 60° S between 45° E and 160° E, excluding the French sector of Terra Adélie (Adélie Land).

Chapter 4: Island

1 Sir Douglas Mawson, *Macquarie Island: A Sanctuary for Australasian Sub-Antarctic Fauna: a lecture*, delivered before the Royal Geographical Society of Australasia (South Australian Branch) Incorporated, Adelaide, 12 September 1919, p.2.

2 Biologist Mary Gillham first used the term 'albatross latitudes' to describe the belt of westerly winds between latitudes 40° S and 60° S, in her book: Mary E. Gillham, *Sub-Antarctic Sanctuary: Summertime on Macquarie Island*, Wellington: A.H. and A.W. Reed, 1967, p.20. See Bernadette Hince, *The Antarctic Dictionary: A Complete Guide to Antarctic English*, Canberra: CSIRO Publishing, 2000, p.3.

3 Wandering albatrosses feed on fish, cephalopods, jellyfish and crustaceans, as well as dead penguins or seals, their foraging trips lasting up to 50 days at a time. They have a 50-year life span and, at maturity, the adult birds will try to breed every two years.

ANARE has been monitoring albatross populations since the 1950s and collecting detailed data for 20 years.

4 Jaimie Cleeland, interview with author, Canberra, 17 September 2019. Dr Cleeland participated in the Macquarie Island ANARE program for three consecutive summers of 2011–2012, 2012–2013 and 2013–2014.

5 Cleeland, interview with author, Canberra, 17 September 2019.

6 Gordon Elliott Fogg, *A History of Antarctic Science*, Cambridge: Cambridge University Press, 1992, p.58.

7 Cited in 'Sydney', *The Sydney Gazette and New South Wales Advertiser*, 13 December 1822, p.2.

8 *Macquarie Island Nature Reserve and World Heritage Area Management Plan 2006*, Parks and Wildlife Service, Hobart, 2006, Appendix 1, p.146.

9 Captain J.K. Davis, 'The ship's story', in Mawson, *The Home of the Blizzard*, vol.1, p.30. The National Library of Australia holds several manuscripts from Mawson's Macquarie Island party, including the official handwritten diaries of Leslie Blake and Harold Hamilton (MS 6650).

10 The Macquarie Island party consisted of George Ainsworth (leader and meteorologist), Leslie Blake (cartographer and geologist), Harold Hamilton (biologist), Charles Sandell (wireless operator and mechanic) and Arthur Sawyer (wireless operator). The observations were continued by others until the station was closed in 1915. See H.J. Gibbney, 'Ainsworth, George Frederick (1878–1950)', *Australian Dictionary of Biography*, vol.7, Carlton: Melbourne University Press, 1979.

11 Mawson, *Macquarie Island: A Sanctuary for Australasian Sub-Antarctic Fauna*, p.2.

12 Mawson, *The Home of the Blizzard*, vol.1, pp.41–42.

13 World Heritage Centre, 'Macquarie Island', World Heritage List, Paris: United Nations Educational, Scientific and Cultural Organization.

14 Isobel Bennett, *Shores of Macquarie Island*, Adelaide: Rigby, 1971, pp.22, 33, 37, 68–69. Her papers are held in the National Library of Australia Manuscripts Collection. See also 'Dr Isobel Bennett (1909–2008), marine biologist', interview by Nessy Allen, Australian Academy of Science, 2000.

15 Tasmanian Government, Department of Primary Industries, Parks, Water and Environment, *Macquarie Island Pest Eradication Project*, Evaluation Report, August 2014.

16 John Stanley Cumpston, *Macquarie Island*, Australian Antarctic Division, Canberra: Department

of External Affairs (Australia), 1968.

17 'Mr Hatch's lecture', *Evening Star* (Dunedin), no.9807, 21 September 1895, p.4; 'Macquarie Islands', *Tasmanian News* (Hobart), 13 June 1895, p.2.

18 Douglas Mawson, *Macquarie Island: Its Geography and Geology*, Australasian Antarctic Expedition 1911–1914, Scientific Reports, series A, vol.5, Sydney: Government Printer, 1943, p.15.

19 Leslie Blake, cited in Douglas Mawson, *Geographical Narrative and Cartography*, Australasian Antarctic Expedition 1911–1914, Scientific Reports, series A, vol.1, part 1, Sydney: Government Printer, 1942, p.281.

20 'Sinking a Small Fortune: Joseph Hatch and the Oiling Industry', *The Sealers' Shanty: A Journal of Sealers' Stories*, vol.9: 1889–1919, pp.1–4.

21 Mawson, *Macquarie Island: A Sanctuary for Australasian Sub-Antarctic Fauna*, pp.1, 7–8.

22 The island continues to be administered by Tasmania as a State Reserve and is now recognised as a place of World Heritage significance.

23 Patrick G. Quilty and Graeme E. Wheller, 'Heard Island and the McDonald Islands: A Window into the Kerguelen Plateau', *Papers and Proceedings of the Royal Society of Tasmania*, vol.133, no.2, 2000, pp.1–12.

24 John Béchervaise, Journal, 21 February and 22 February 1953, papers of John Béchervaise, National Library of Australia MS 7972.

25 Sir C. Wyville Thomson, *The Voyage of the 'Challenger': The Atlantic, a Preliminary Account of the General Results of the Exploring Voyage of HMS 'Challenger' During the Year 1873 and the Early Part of the Year 1876*, London: Macmillan and Co, 1877, p.220.

26 Arthur Scholes, *Fourteen Men: The Story of the Australian Antarctic Expedition to Heard Island*, Melbourne: F.W. Cheshire, 1949, p.30.

27 Australia's other active volcano is on the nearby McDonald Islands. For an excellent history of six subantarctic islands, including Macquarie and Heard islands, see Bernadette Hince, 'The Teeth of the Wind: An Environmental History of Subantarctic Islands', PhD thesis, Australian National University, 2005.

28 Cabinet Submission, For Full Cabinet, Top Secret, Agendum No 1275E, p.2, cited in Marie Kawaja, 'The Politics and Diplomacy of the Australian Antarctic, 1901–1945', PhD thesis, Australian National University, May 2010, p.18, Bernadette Hince, 'The Cultural History of the Eastern Hemisphere

Subantarctic Islands', e. ISIC4P Hince.pdf, sicri-network.org; Hince, 'The Teeth of the Wind'.

29 Britain's argument for territorial control of Heard Island rested on the claim that a British sealer had been the first to sight the island, in 1833. Australia's claim in 1947 served to confirm its tenure under the British Crown. See Hince, 'The Teeth of the Wind', pp.117–119.

30 *Antarctica 1948* was released for public viewing in 1949. See National Film and Sound Archive of Australia, *Antarctica 1948*, 1949, aso.gov.au/titles/documentaries/antarctica–1948.

31 Scholes, *Fourteen Men*, p.30.

32 Scholes, *Fourteen Men*, p.62.

33 Scholes, *Fourteen Men*, p.65.

34 Scholes, *Fourteen Men*, pp.71–72.

35 Chief engineer, cited in Scholes, *Fourteen Men*, p.74.

36 Scholes, *Fourteen Men*, p.75; Graeme M. Budd, 'Australian Exploration of Heard Island, 1947–1971', *Polar Record*, vol.43, no.225, 2007, pp.97–123.

37 Dr Phillip Law succeeded Campbell and served as Leader of ANARE and Director of the Antarctic Division from 1949 to 1966.

Chapter 5: Territory

1   Mawson, *The Home of the Blizzard*, vol.2, p.259.

2   Geoff Mosley, *Saving the Antarctic Wilderness: The Pivotal Role in its Complete Protection*, Canterbury: Envirobook, 2009, p.3.

3   P. Quilty, 'Introducing Antarctica', *Issues: 'Why Does the World need Antarctica?'*, Melbourne: Australian Council for Education Research, March 1995, pp.9–10; Tim Bowden, *The Silence Calling: Australians in Antarctica 1947–97*, The ANARE Jubilee History, Sydney: Allen & Unwin, 1997, p.249.

4   Alessandro Antonello, 'Glaciological Bodies: Australian Visions of the Antarctic Ice Sheet', *International Review of Environmental History*, vol.4, no.1, 2018, pp.132. Given the lack of glaciologists in Australia, Law turned to physicists and mathematicians and trained them as glaciologists. See Bowden, *The Silence Calling*, p.266.

5   Phillip Law, *Antarctic Odyssey*, Melbourne: Heinemann, 1983, p.175.

6   'Empire builder in Antarctica', *The Age* (Melbourne), 9 February 1954, p.1. The voyage marked the beginning of a 34 year association between the Danish shipping company J. Lauritzen and Co. of Copenhagen and Australia's Antarctic program.

7 Bowden, *The Silence Calling*, pp.103–104.

8 Law, *Antarctic Odyssey*, p.16.

9 Dovers had spent a year as surveyor with the first ANARE party on Heard Island in 1948 and five months on Macquarie Island.

10 Law, *Antarctic Odyssey*, p.104.

11 Louis Edward Macy, Papers related to Antarctica, 1947–1984, Mitchell Library, State Library of New South Wales, ML MSS 5343

12 Law, *Antarctic Odyssey*, pp.174–175. A large seabird breeding site, Scullin Monolith was later given legal protection, along with Rookery Islands and another rocky coastal outcrop, Murray Monolith.

13 Friends of Nella Dan website, nelladan.org/en/about-the-ship.

14 Sarah Laverick, *Through Ice & Fire: The Adventures, Science and People behind Australia's Famous Icebreaker* Aurora Australis, Sydney: Pan Macmillan, 2019; Australian Government, 'RSV *Aurora Australis* 1989–2020', Australian Antarctic Program, antarctica.gov.au/about-antarctica/history/transportation/shipping/aurora-australis.

15 Australian Antarctic Division, 'RSV *Nuyina* Arrives in Hobart', antarctica.gov.au/nuyina/arrival.

16 Sir Hubert, from South Australia, made several experimental flights in the Arctic and, in 1928, completed the first trans-Arctic flight from Alaska to Spitzbergen. He received a knighthood in recognition of his achievements. 'Conquest of the Antarctic: Two Expeditions to Explore Frozen South', *The Mercury* (Hobart), 27 December 1933, p.6.

17 The rock cairn and a canister containing a facsimile of the proclamation is located at 68°22' S, 78°32' E.

18 Phillip Law, *Australia and the Antarctic*, The John Murtagh Macrossan Memorial Lectures, 1960, St Lucia: University of Queensland Press, 1962, pp.5 and 7.

19 Phillip Law, 'Australian National Antarctic Research Expedition, 1955', *The Geographical Journal*, vol.122, no.1, 1956, p.34 [31–39].

20 Phillip Law, 'Australian Coastal Exploration in Antarctica', *The Geographical Journal*, vol.124, no.2, 1958, p.161 [151–162].

21 Australian Antarctic Division, 'Living at Mawson', antarctica.gov.au/antarctic-operations/stations/mawson/living.

22 Australian Antarctic Division, 'Mawson Station: A Brief History', antarctica.gov.au/about-antarctica/history/stations/mawson.

23 Australian Antarctic Division, 'Davis Station: A Brief History', antarctica.gov.au/about-antarctica/history/stations/davis.

24 Australian Antarctic Division, 'Casey Station: A Brief History', antarctica.gov.au/about-antarctica/history/stations/casey.

25 Australian Antarctic Division, 'Macquarie Island Station: A Brief History', antarctica.gov.au/about-antarctica/history/stations/macquarie-island.

26 Law, *Antarctic Odyssey*, p.267.

27 Secretariat of the Antarctic Treaty website, ats.aq/index_e.html; Alessandro Antonello, *The Greening of Antarctica: Assembling an International Environment*, Oxford: Oxford University Press, 2019, p.9.

28 C. Collis, 'The Proclamation Island Moment: Making Antarctica Australian', *Law Text Culture*, vol.8, article 3, 2004, ro.uow.edu.au/ltc/vol8/iss1/3; The Antarctic Treaty, ats.aq/e/antarctictreaty.html; Bowden, *The Silence Calling*, p.178.

29 Igor Krupnik et al. (eds), *Understanding Earth's Polar Challenges: International Polar Year 2007–2008*, Alberta: International Polar Year Joint Committee, 2011, scar.org/library/scar-publications/occasional-publications/3505-understanding-earth-s-polar-challenges/file; 'Australia's Contribution to the International Polar Year', *Australian Antarctic Magazine*, issue 16, 2009.

30 Vin Morgan, cited in Linda Clark and Elspeth Wishart, *66 South: Tales from an Antarctic Station*, Launceston: Queen Victoria Museum and Art Gallery, 1993, p.27.

31 Keith Gooley, interview with author, 16 November 2019.

32 Clark and Wishart, *66 South*, pp.29–30.

33 Wilkes is now entombed in the ice, apart from occasional summer thaws when the remains of a station built for earlier times can still be seen. Visitors from Casey still use the old transmitter building, known as the 'Wilkes Hilton', for field accommodation. See Australian Antarctic Division, 'Wilkes Station', antarctica.gov.au/about-antarctica/history/stations/wilkes.

34 Gooley, interview with author, 16 November 2019.

35 Phillip Law, cited in Clark and Wishart, *66 South*, p.28.

36 Charlie Weir, cited in Clark and Wishart, *66 South*, p.52.

37 Stephen Murray-Smith, *Sitting on Penguins: People and Politics in Australian Antarctica*, Surry Hills: Hutchinson Australia, 1988, pp.88–90.

38 Graeme Manning, cited in Clark and Wishart, *66 South*, p.31; Bowden, *The Silence Calling*, pp.347–348.

39 'Tubby' Seton, cited in Clark and Wishart, *66 South*, p.32. Also see Alan White, cited in Clark and Wishart, *66 South*, p.31.

40 Australian Antarctic Division, 'Australia's Antarctic Buildings: AANBUS', Australian Government, antarctica.gov.au/about-antarctica/history/stations/history-of-australias-antarctic-buildings/beginnings-of-aanbus.

41 See, for example, Dale Main and Joan Russel, both cited in Clark and Wishart, *66 South*, p.33.

42 Murray-Smith, *Sitting on Penguins*, pp.90–91.

Chapter 6: Station

1   John Béchervaise, *Blizzard and Fire: A Year at Mawson, Antarctica*, Sydney: Angus and Robertson, 1963, p.143.

2   Béchervaise, *Blizzard and Fire*, p.55.

3   Béchervaise, *Blizzard and Fire*, p.xi.

4   Peter Keage, 'Obituary: John Béchervaise', *Independent*, 2 September 1998.

5   Law, *Australia and the Antarctic*, p.5.

6   Béchervaise, *Blizzard and Fire*, pp.48–49.

7   Béchervaise, *Blizzard and Fire*, p.62.

8   Béchervaise, *Blizzard and Fire*, pp.64–67.

9   Béchervaise, *Blizzard and Fire*, p.142.

10 Béchervaise, *Blizzard and Fire*, p.121.

11 Béchervaise, *Blizzard and Fire*, p.123.

12 Béchervaise, *Blizzard and Fire*, pp.125 and 127.

13 Béchervaise, *Blizzard and Fire*, p.127.

14 Béchervaise, *Blizzard and Fire*, pp.137–141.

15 Béchervaise, *Blizzard and Fire*, p.217.

16 Béchervaise, *Blizzard and Fire*, p.227.

17 Béchervaise, *Blizzard and Fire*, p.129.

18 Béchervaise, *Blizzard and Fire*, pp.238–239.

19 Béchervaise, *Blizzard and Fire*, p.241.

20 John Béchervaise, *Australia and Antarctica*, Lane Cove: Nelson Doubleday Pty Ltd, 1967, pp.5–7, 11. Béchervaise led three wintering expeditions to Antarctica and published several books on the polar regions. He was awarded the Queen's Polar Medal and MBE for his services to Antarctic science.

21 Béchervaise, *Blizzard and Fire*, p.230.

22 Elizabeth Leane et al., 'Beyond the Heroic Stereotype: Sidney Jeffryes and the Mythologising of Australian Antarctic History', *Australian Humanities Review*, vol 64, pp.1–28; 'Sydney Jeffryes remembered', *Australian Antarctic Magazine*, issue 35, December 2018, antarctica.gov.au/magazine/

issue-35-december-2018/in-brief/sidney-jeffryes-remembered.

23 Rod Mackenzie, cited in Clark and Wishart, *66 South*, p.89.

24 See Hince, *The Antarctic Dictionary*, p.389.

25 Morgyn Phillips, 'Extreme film and sound', National Film and Sound Archive of Australia, nfsa.gov.au/latest/extreme-film-and-sound-stories-antarctica; Wilkes Station History, sites.google.com/site/wilkesstationhistory.

26 Joan Saxton, cited in Clark and Wishart, *66 South*, p.91.

27 Mrs Rob, cited in Clark and Wishart, *66 South*, p.91.

28 Raymond Priestley, cited in Griffiths, *Slicing the Silence*, p.172.

29 Graeme Manning and Kim Hill, cited in Clark and Wishart, *66 South*, pp.80–81.

30 Graeme Manning, cited in Clark and Wishart, *66 South*, p.82.

31 Richard Penney, cited in Clark and Wishart, *66 South*, p.82.

32 Rod Mackenzie, cited in Clark and Wishart, *66 South*, p.82.

33 Rod Mackenzie, cited in Clark and Wishart, *66 South*, p.83.

34 Alan White, cited in Clark and Wishart, *66 South*, p.79.

35 Station yearbooks and station leaders' reports represent an important archive of life on the Antarctic station and a valuable source of information for expeditioners in subsequent years.

36 Elizabeth Parer-Cook (with Neil Roberts), *Mawson 1972: Reflections 40 Years On*, Brighton: Zaurora Books, 2012, p.45.

37 Parer-Cook and Roberts, *Mawson 1972*, p.54.

38 Parer-Cook and Roberts, *Mawson 1972*, p.52.

39 Nel Law produced a series of paintings from that expedition and designed an emblem for ANARE which was subsequently simplified. See the National Library of Australia Map Collection: MAP SCAR Coll. drawer 3, folder 4, map 44. The simplified symbol featured a leopard seal, which was in use until 1985, when it was replaced by a new 'globe' logo featuring Australia and Antarctica. Law's diaries, poems and sketches are held by the Australian Science and Technology Heritage Centre at the University of Melbourne, series 18: Nel Law. 'Ian Toohill Talks About Phil Law', ANARE Club website, anareclub.org.

40 Shane Hill, cited in Clark and Wishart, *66 South*, pp.38–39. The earliest known woman was Jeanne

Baré (Baret), a valet and assistant naturalist with Bougainville's French expedition to the Falkland Islands aboard *L'Etoile* in 1766–1767. See Elizabeth Chipman, *Women on the Ice: A History of Women in the Far South*, Carlton: Melbourne University Press, 1986, p.26.

41 Robin Burns, 'Women in Antarctica: Sharing This Life-changing Experience', Fourth Phillip Law Lecture, Hobart, 18 June 2005, Hobart: Department of State Growth, stategrowth.tas.gov.au/__data/assets/pdf_file/0013/2092/Dr_Robin_Burns_Lecture_-_No._4.pdf, p.3.

42 Morton Henry Moyes, Diary, Mitchell Library, State Library of New South Wales, ML MSS 388/1, CY 3660; M. Moyes, as told to G. Dovers and D. Niland, 'Season in solitary', *Walkabout*, vol.30, no.10, 1964, pp.21 and 23.

43 Louise Holliday, cited in 'I Was the First Woman to Spend a Winter in Antarctica', *Woman's Day* (Australia), 5 November 2018; Keith Finlay, 'My First Strange Weeks In Antarctica', *The Australian Women's Weekly*, 18 February 1981, p.13.

44 Bowden, *The Silence Calling*, p.453.

45 For a fuller discussion of women's participation in ANARE life, see Elizabeth Chipman, *Women on the Ice: A History of Women in the Far South*, Carlton: Melbourne University Press, 1986; and Bowden, *The Silence Calling*, pp.443–468.

46 Rear Admiral George Dufek, cited in Burns, *Just Tell Them I Survived!*, p.23.

47 See, for example, Jesse Blackadder, 'Frozen Voices: Women, Silence and Antarctica', in Bernadette Hince et al. (eds), *Antarctica: Music, Sounds and Cultural Connections*, Canberra: ANU Press, 2015, p.170 [169–177].

48 Since then, women have assumed a variety of professional roles, including as station leaders, scientists, technicians and chefs, as well as officers and crew members aboard the resupply vessels. See, for example, Aspa Sarris, 'Antarctic Station Life: The First 15 Years of Mixed Expeditions to the Antarctic', *Acta Astronautica*, vol.131, 2017, pp.50–54.

49 Diana Patterson, *The Ice Beneath My Feet: My Year in Antarctica*, Sydney: ABC Books, 2010, p.146.

50 Diana Patterson, interview with Tim Bowden, Sydney, 16 March 1996.

51 Mawson, *The Home of the Blizzard*, vol.2, pp.2–3.

52 Wilkes Station History, sites.google.com/site/wilkesstationhistory.

53 Dale Main, cited in Clark and Wishart, *66 South*, p.84.

54 Australian Antarctic Division, 'This Week at Casey: 16 August 2019', antarctica.gov.au/news/stations/Casey/2019/this-week-at-casey-16-august-2019.

55 Denise Allen, interview with author, 15 November 2019. All four of Australia's permanent stations (three on the continent, and Macquarie Island) record meteorological data each day for transmission to the Australian Bureau of Meteorology.

56 Denise Allen, interview with author, 15 November 2019.

57 Denise Allen, interview with Bruce Watson in Melbourne, 13–14 June 2011, Australian Antarctic culture oral history project, National Library of Australia, ORAL TRC 6275/2.

58 Allen, interview with author, 15 November 2019.

59 Allen, interview with author, 15 November 2019.

60 Allen, interview with author, 15 November 2019.

Chapter 7: Inland

1 Bernacchi, *To the South Polar Regions*, p.250.

2 Ken G. McCracken et al., 'The Stimulus of New Technologies', in Harvey J. Marchant et al. (eds), *Australian Antarctic Science: The First 50 Years of ANARE*, Kingston: Australian Antarctic Division, 2002, pp.27–28. The Weasel was used extensively for military operations by the US army, and subsequently deployed in the Arctic before being sold off in the 1950s for civilian use. The Australian Antarctic Division (AAD) holds one of its Weasels in its heritage collection in Kingston, Tasmania. The AAD eventually replaced the Weasel with modern tracked vehicles for long-distance fieldwork, including the Swedish dual-cab Hägglunds and Nodwell and Canadian Foremost Pioneer. See 'Journey of Discovery for Antarctic Heritage Specialists', *Australian Antarctic Magazine*, issue 27, December 2014, antarctica.gov.au/magazine/2011-2015/issue-27-december-2014/history/journey-of-discovery-for-antarctic-heritage-specialists.

3 William F. Budd, 'The Antarctic Ice Sheet', in Marchant et al. (eds), *Australian Antarctic Science*, pp.317–318.

4 Law, *Antarctic Odyssey*, p.270.

5 Mawson, *The Home of the Blizzard*, vol.1, p.109.

6 William Budd, cited in Clark and Wishart, *66 South*, p.54.

7 Rod Mackenzie, cited in Clark and Wishart, *66 South*, p.54; Parer-Cook and Roberts, *Mawson 1972*, p.54.

8 Wendy Pyper, 'Antarctic Place Names Go to the

Dogs', *Australian Antarctic Magazine*, issue 33, December 2017, antarctica.gov.au/magazine/issue-33-december-2017/history/antarctic-place-names-go-to-the-dogs; Australian Antarctic Division, 'Gadget & Ginger Land Safely After Historic Flight to Antarctica', 30 December 2004, antarctica.gov.au/news/2004/gadget-and-ginger-land-safely.

9   See, for example, Jesse Blackadder, *Stay: The Last Dog in Antarctica*, Sydney: HarperCollins Publishers Australia, 2013; *Stay the Dog: The Last Canine in Antarctica*, The Australasian Antarctic Expedition 2013–2014, 23 November 2015, youtube.com/watch?v=G-4xVCZxmGs.

10 Neville Collins, cited in Clark and Wishart, *66 South*, pp.49 and 52. Collins was a winter expeditioner on several occasions, including Mawson (1957, 1960), Wilkes (1962) and Amery Ice Shelf (1968). Apart from mechanical repairs, one of the more arduous chores was preheating the engines for up to seven hours before they could be started.

11 Collins, cited in Clark and Wishart, *66 South*, p.49.

12 Griffiths, *Slicing the Silence*, p.103; Robert Thomson, *The Coldest Place on Earth*, Wellington: A.H. & A.W. Reed, 1969; Clark and Wishart, *66 South*, p.70–74.

13 The Vostok team measured an ice depth of 4,880 metres, recorded 760 kilometres from Wilkes. 'Wilkes–Vostok Seismic Traverse, 1962–63', from a note by R.B. Thomson and press releases, *Polar Record*, vol.12, no.77, 1964, pp.192–193; transcript from *Vostok 900* [film], Australian Antarctic Division archives, antarctica.gov.au/about-antarctica/history/exploration-and-expeditions/modern-expeditions/stories-from-the-archives/steak-and-onions-vostok-style#v185514. See also Phillip Law in Marchant et al. (eds), *Australian Antarctic Science*, p.18.

14 Neville Collins, interview with Tim Bowden, 14 July 1994; Tim Bowden, *The Silence Calling*, pp.253–256.

15 Bowden, *The Silence Calling*, p.256.

16 Murray-Smith, *Sitting on Penguins*, pp.133–134; Wendy Pyper, 'Traversing Antarctica', *Australian Antarctic Magazine*, issue 33, December 2017, antarctica.gov.au/magazine/2016-2020/issue-33-december-2017/operations/traversing-antarctica.

17 Keith Gooley, interview with author, 16 November 2019. The IPS is now known as Space Weather Services.

18 Gooley, interview with author, 16 November 2019.

19 Gooley, interview with author, 16 November 2019.

20 Phillip Law, 'The IGY in Antarctica', *The Australian Journal of Science*, vol.21, no.9, 1959, p.292 [285–294]; Budd, 'The Antarctic Ice Sheet', in Marchant et

al. (eds), *Australian Antarctic Science*, pp.309–390.

21 Australian Antarctic Division, 'Ice Sheets', updated 13 May 2011, antarctica.gov.au/about antarctica/ice-and-atmosphere/ice-sheet.

22 Patrick Quilty, cited in 'Antarctic Community Mourns Loss of Science Leader', *Australian Antarctic Magazine*, issue 35, December 2018.

23 Murray-Smith, *Sitting on Penguins*, p.175; R. Ewan Fordyce et al., '*Australodelphis mirus*, a Bizarre New Toothless Ziphiid-like Fossil Dolphin (Cetacea: Delphinidae) from the Pliocene of Vestfold Hills, East Antarctica', *Antarctic Science*, vol.14, no.1, 2001, pp.37–54.

24 Murray-Smith, *Sitting on Penguins*, p.177.

25 Bowden, *The Silence Calling*, p.499.

26 Murray-Smith, *Sitting on Penguins*, p.169. In 1987, the Antarctic Treaty Committee recognised Marine Plain as a Site of Special Scientific Interest, and it was redesignated in 1996 as Antarctic Specially Protected Area No.143, 'a major Antarctic terrestrial ice-free ecosystem with outstanding fossil fauna and rare geological features'. National Science Foundation, nsf.gov/geo/opp/antarct/aca/nsf01151/aca2_spa143.pdf.

27 Michael Pearson, 'Sledges and Sledging in Polar Regions', *Polar Record*, vol.31, issue 1976, 1995, pp.3–24.

28 McCracken et al., 'The Stimulus of New Technologies', in Marchant et al. (eds), *Australian Antarctic Science*, pp.27–28.

29 The Australian Government contracted two CASA 212–400 fixed-wing aircraft to provide transport and field support for the Australian Antarctic Program. See The Hon. Dr Sharman Stone MP, 'From Huskies to Otters: A New Antarctic Era Arrives', *Media release*, SS03/121, Australian Parliament, 26 November 2003; Australian Antarctic Program, 'Davis and Mawson Changeover', 3 November 2003, antarctica.gov.au/news/2003/davis-and-mawson-changeover.

30 John Béchervaise, *Blizzard and Fire*, p.101.

Chapter 8: Ocean

1   Tony Press, 'In Deep: Australian Research in the Southern Ocean', *Australian Antarctic Magazine*, no.4, spring 2002, pp.1–2.

2   Charles Wyville Thomson, 'Hydrographic Instructions to Captain G. S. Nares, HMS *Challenger*, in Thomas Henry Tizard et al., *Narrative of the Cruise of HMS* Challenger, *with a General Account of*

the *Scientific Results of the Expedition*, vol.1 of Great Britain, *Challenger* Office, Report on the Scientific Results of the Voyage of HMS *Challenger* during the Years 1873–76 under the Command of Captain George S. Nares, RN, FRS and the Late Captain Frank Tourle Thomson, RN, Edinburgh: Neill, 1885, part 1, p.34 [34–39]. The British *Challenger* expedition spent four years circumnavigating the world investigating marine life, measuring the depth, temperature and salinity of the oceans and determining the nature of the sea floor.

3   Davis, *High Latitude*, p.152.

4   Captain J.K. Davis, 'The Ship's Story', in Mawson, *The Home of the Blizzard*, vol.2, p.19.

5   Mawson, *The Home of the Blizzard*, vol.2, Appendix II, Scientific work.

6   Stephen R. Rintoul et al., 'The Antarctic Circumpolar Current System', *Ocean Circulation and Climate*, 2001, pp.271–302, epic.awi.de/id/eprint/2649/1/Rin8888b.pdf.

7   Wallace S. Broecker, *The Great Ocean Conveyor. Discovering the Trigger for Abrupt Climate Change*, Princeton: Princeton University Press, 2010; Wallace S. Broecker, 'Unpleasant Surprises in the Greenhouse?', *Nature*, vol.328, no.6126, 9 July 1987, pp.123–126; Tom Griffiths, 'Ice Cores and Climate Change', in Libby Robin et al. (eds), *The Future of Nature: Documents of Global Change*, New Haven: Yale University Press, 2013; Jo Chandler, *Feeling the Heat*, Melbourne: Melbourne University Publishing, 2011.

8   Susan Barr and Cornelia Lüdecke (eds), *The History of the International Polar Years (IPYs)*, Berlin: Springer-Verlag, 2010.

9   'Southern Ocean Winds Open Window to the Deep Sea', *Science Daily*, 17 March 2010, sciencedaily.com/releases/2010/03/100315103820.htm.

10 See, for example, Nathan Bindoff et al., *Position Analysis: Climate Change and the Southern Ocean*, Hobart: Antarctic Climate and Ecosystems Cooperative Research Centre, 2011, pp.3–4.

11 Narissa Bax, interview with author, 14 November 2019. The project, led by the AAD's Principal Research Scientist Dr Andrew Constable, aimed to assess the vulnerability to fishing gear of seafloor habitats and associated invertebrate communities. The voyage was conducted both for research and resupplying Australia's Antarctic stations.

12 Bax, interview with author, 14 November 2019.

13 *Census of Marine Life*, coml.org/about-census.

14 Wendy Pyper, 'Southern Ocean Marine Life in Focus',

Australian Antarctic Magazine, issue 18, 2010.

15 Traditionally, everyone participates in a work roster to clean the station buildings and do 'slushy' duty in the kitchen, while skilled tradespeople undertake maintenance work.

16 Bax, interview with author, 14 November 2019.

17 Bax, interview with author, 14 November 2019.

18 David Barnes, 'Antarctic Seabed Carbon Capture Change' (ASCCC), British Antarctic Survey, bas.ac.uk/project/antarcticseabedcarboncapturechange.

19 The process results in a major carbon reservoir, referred to as 'blue carbon', the most rapidly increasing carbon store and negative feedback on climate on the planet. Narissa Bax et al., 'Perspective: Increasing Blue Carbon around Antarctica is an Ecosystem Service of Considerable Societal and Economic Value Worth Protecting', *Global Change Biology*, vol.27, no.1, 2021, pp.5–12; Tony Press, 'In Deep: Australian Research in the Southern Ocean', *Australian Antarctic Magazine*, no.4, spring 2002, pp.1–7.

20 McCann, *Wild Sea*, p.86.

21 Bernacchi, *To the South Polar Regions*, p.31.

22 Bernacchi, *To the South Polar Regions*, p.39.

23 Law, *Antarctic Odyssey*, p.16.

24 Australian Antarctic Division, 'Sea Ice', antarctica.gov.au/about-antarctica/environment/sea-ice.

25 Tracey Rogers, cited in Robin Burns, *Just Tell Them I Survived!*, p.164.

26 Rogers, cited in Burns, *Just Tell Them I Survived!*, p.164.

27 John Béchervaise, *Blizzard and Fire*, pp.28–41.

28 See Hince, *The Antarctic Dictionary*, p.14, for a definition of the 'A Factor'.

29 Australian Antarctic Program, 'The A Factor in the Mawson Fly-off (2006)', antarctica.gov.au/about-antarctica/history/exploration-and-expeditions/expeditioner-stories/the-a-factor-in-the-mawson-fly-off.

30 Barbara Wienecke, interview with author, 30 September 2019.

31 Apsley Cherry-Garrard, *The Worst Journey in the World*, p.240.

32 During her first visit to Antarctica, Wienecke and a companion spent seven months in the field over winter, accommodated in a small hut on a tiny rock in the sea ice known as Macey Island. She was awarded the Australian Antarctic Medal in 2013 for her work with Emperor penguins.

33 Wienecke, interview with author, 30 September 2019.

34 Australian Antarctic Division, 'Finding Emperor Penguin Colonies', antarctica.gov.au/about-

antarctica/animals/penguins/emperor-penguins/
finding-emperor-penguin-colonies.

35 Wienecke, interview with author, 30 September 2019.

36 Wienecke, interview with author, 30 September 2019.

37 Stephen Nicol, *The Curious Life of Krill: A
Conservation Story from the Bottom of the World*,
Washington DC: Island Press, 2018, p.11; also see
McCann, *Wild Sea*, pp.186–187.

38 Commission for the Conservation of Antarctic
Marine Living Resources website, www.ccamlr.org.

39 Wienecke, interview with author, 30 September 2019.

40 Wienecke, interview with author, 30 September 2019.

Chapter 9: Wilderness

1   Sidney Nolan, cited in Rowena Wiseman et al.
(eds), *Sidney Nolan: Antarctic Journey*, Melbourne:
Mornington Peninsula Regional Gallery, 2006. The
National Library of Australia holds the papers of
Sidney Nolan, diaries and notebooks, 1949–1991, in
its Manuscript Collection.

2   Francis Spufford, 'The Uses of Antarctica: Roles
for the Southern Continent in Twentieth-century
Culture', in Ralph Crane et al. (eds), *Imagining
Antarctica: Cultural Perspectives on the Southern
Continent*, Hobart: Quintus Publishing, 2011,
p.17 [17–30].

3   Graham Richardson, Minister for the Arts, Sport, the
Environment, Tourism and Territories, 'Foreword',
in Commonwealth of Australia, *Antarctic Journey:
Three Artists in Antarctica*, Canberra: Australian
Government Publishing Service, 1988, p.iii. The
voyage included a visit to Heard Island, the Scullin
Monolith, Davis station and Vestfold Hills. All three
produced a prolific output of artwork, despite Bea
Maddock's leg injury, sustained at Heard Island.

4   Murray-Smith, *Sitting on Penguins*, p.225. The results
of this voyage were published in a series of articles in
The Australian during the visit, and subsequently in
the book *Sitting on Penguins* (1988). His biographer,
historian Ken Inglis, attributes Murray-Smith with
helping to bring about some of the changes that
sought to improve the way that Australia conducted
its activities in Antarctica. See K.S. Inglis, 'Murray-
Smith, Stephen (1922–1988)', *Australian Dictionary
of Biography*, vol.18, Carlton: Melbourne University
Press, 2012.

5   Murray-Smith, *Sitting on Penguins*, p.225.

6   Murray-Smith, *Sitting on Penguins*, p.99.

7   Murray-Smith, *Sitting on Penguins*, p.226.

8   Elizabeth Leane, 'Introduction', in Crane et al. (eds),
Imagining Antarctica, p.12.

9   Wiseman et al. (eds), *Sidney Nolan*, p.1.

10 Wiseman et al. (eds), *Sidney Nolan*, p.3. Antarctic
photography included black-and-white and colour
images as well as cinematographic footage and
movies. In the early 1970s, Dave Parer, nephew of
renowned Australian cinematographer Damien Peter
Parer, spent two summers studying Antarctic physics
at Mawson station and, with Elizabeth Parer-Cook,
made two nature documentaries there, *Antarctic
Winter* and *Antarctic Summer*.

11 The Frank Hurley collection in the National
Library of Australia contains more than 10,000
photographic negatives and lantern slides purchased
from the Hurley family, and includes more than
250 photographic prints and negatives taken in
Antarctica and the subantarctic islands.

12 Alasdair McGregor, *Frank Hurley: A Photographer's
Life*, Camberwell: Viking, 2004, p.35.

13 McLean, diary, n.d., cited in McGregor, *Frank Hurley*,
p.42.

14 McGregor, *Frank Hurley*, opp.p.88. McGregor
notes that few of Hurley's 'beautiful studies' of the
Antarctic landscape were ever seen.

15 Frank Worsley, diary, 24 January 1915, cited in
McGregor, *Frank Hurley*, p.99.

16 Frank Hurley, diary, 27 August 1915, cited in
McGregor, *Frank Hurley*, p.111.

17 Frank Hurley, *Shackleton's Argonauts: A Saga of the
Antarctic Ice-Packs*, Angus and Robertson, Sydney,
1948, p.140.

18 Hurley, *Shackleton's Argonauts*, p.127.

19 Tom Griffiths, 'Debunking Mawson', *Inside Story*,
3 December 2013, insidestory.org.au/debunking-
mawson.

20 Mawson's Huts Replica Museum, mawsons-huts.org.
au/replica-museum.

21 Griffiths, *Slicing the Silence*, pp.330–331.

22 'Polar Exploration. Lecture By Dr Mawson', *Evening
Journal* (Adelaide), 27 July 1910, p.2.

23 'Exhibition May Be Prolonged', *The Register*
(Adelaide), 28 April 1910, p.6.

24 Mawson to Hurley, 16 November 1916, MAC 6DM,
cited in McGregor, *Frank Hurley*, p.145.

25 His photographs appeared in newspapers in Britain
and Australia, including a montage titled 'Mawson's
Expedition to the Ice-bound South', in *The Observer*
(Adelaide), 7 March 1914, p.31.

26 Hurley's films, including *The Home of the Blizzard:
Life in the Antarctic* (The Official Film of the Mawson

Antarctic Expedition (c. 1916), *Endurance* (1917) and *Siege of the South* (1931), are held in the National Film and Sound Archive of Australia collection. He travelled to Antarctica six times during his career, in addition to other extensive journeys and as official photographer with the AIF in the Middle East during the Second World War. See A.F. Pike, 'Hurley, James Francis (Frank), (1885–1962)', *Australian Dictionary of Biography*, vol.9, Carlton: Melbourne University Press, 1983; Robert Dixon, *Photography, Early Cinema and Colonial Modernity: Frank Hurley's Synchronized Lecture Entertainments*, Anthem Australian Humanities Research Series, London: Anthem Press, 2011.

27 Brigid Hains, *The Ice and the Inland: Mawson, Flynn, and the Myth of the Frontier*, Carlton: Melbourne University Press, 2002.

28 Secretariat of the Antarctic Treaty website, ats.aq/e/antarctictreaty.

29 Between 1960 and 2020, Australia conducted ten such inspections. See Australian Antarctic Division, 'Travel Across Antarctica By Podcast', 30 June 2021, antarctica.gov.au/news/2021/travel-10-000km-across-antarctica-by-podcast.

30 See, for example, Nengye Liu et al., *Governing Marine Living Resources in the Polar Regions*, Cheltenham: Edward Elgar Publishing, 2019.

31 Wiseman et al. (eds), *Sidney Nolan*.

32 Sidney Nolan cited in Wiseman et al., *Sidney Nolan*.

33 Samuel Taylor Coleridge, 'The Rime of the Ancient Mariner', in William Wordsworth and Samuel Taylor Coleridge, *Lyrical Ballads, with a Few Other Poems*, vol.1, London: J. & A. Arch, 1798, pp.15–63; Elizabeth Leane, 'Fictionalizing Antarctica', in Klaus Dodds et al. (eds), *Handbook on the Politics of Antarctica*, Cheltenham: Edward Elgar Publishing Inc., 2017, p.22.

34 Thomas Keneally, 'Captain Scott's Biscuit: An Antarctic Pilgrimage', *Granta*, no.83, 2003, pp.129–144.

35 Monika Schillat, 'Images of Antarctica as Transmitted by Literature' in *Tourism in Antarctica*, Springer Briefs in Geography, New York: Springer, 2016, p.33 [21–39].

36 Harry Black, 'Jumbo Jet to Antarctica', *Aurora: ANARE Club Journal*, Midwinter 1977, p.95.

37 Diane Erceg, 'Explorers of a Different Kind: A History of Antarctic Tourism 1966–2016, PhD thesis, Australian National University, 2017, p.85.

38 'Day-Trippers See a Polar Base', *The Sydney Morning Herald*, 14 February 1977, p.4, cited in Erceg,

'Explorers of a Different Kind', p.88. Other overflights followed.

39 Griffiths, *Slicing the Silence*, pp.332–334.

40 Don McIntyre (with Peter Meredith), *Two Below Zero: A Year Alone in Antarctica*, Terrey Hills: Australian Geographic, 1996. Don and Margie McIntyre later drew on their experience to become motivational speakers and returned to Antarctica as tourist guides.

41 Jarvis began the trek with Russian–Australian John Soukalo and completed it alone in the same 47 days that Mawson had taken. While it was technically a solo trek, a film crew accompanied Jarvis to record it for the documentary film, *The Unforgiving Minute*, made by Film Australia.

42 Tim Jarvis, *Mawson: Life and Death in Antarctica*, Carlton: Miegunyah-Melbourne University Press, 2008, p.139.

43 Margaret Simpson, 'Extreme South: James Castrission and Justin Jones' Antarctic adventure', Sydney: Museum of Applied Arts & Sciences, 2 August 2012, maas.museum/inside-the-collection/2012/08/02/extreme-south-james-castrission-and-justin-jones-antarctic-adventure.

44 Numbers of international tourists increased from 6,700 in 1992–1993 to 45,800 in 2018–2019. Around 11 per cent of those were from Australia. *IAATO Overview of Antarctic Tourism: 2017–18 Season and Preliminary Estimates for 2018–19 Season*, p.4, iaato.org/documents/10157/2398215/IAATO+overview/bc34db24-e1dc-4eab-997a-4401836b7033.

Chapter 10: Environment

1 Law, *Antarctic Odyssey*, p.272.

2 John Heap, 'The Scope of Antarctic Science', in Marchant et al. (eds), *Australian Antarctic Science*, p.14.

3 Griffiths, 'Debunking Mawson', *Inside Story*.

4 Tom Griffiths, 'Australia's Culture of Ice', Ninth Phillip Law Lecture, Hobart, 31 July 2011, *The Phillip Law Lectures*, Antarctic Tasmania and Science Research Development, 2013, vol.3: 2009–2012, p.30 [28–46].

5 Mawson, *The Home of the Blizzard*, vol.1, p.109.

6 Mawson, *The Home of the Blizzard*, vol.1, p.110.

7 'Australia's Antarctica: Portion of a Frozen Continent', *The Kalgoorlie Miner*, 3 May 1939, p.7.

8 Griffiths, *Slicing the Silence*, p.276; Elizabeth Truswell, *A Memory of Ice: The Antarctic Voyage of the* Glomar Challenger, Acton: ANU Press, 2019.

9   Roger Wilson, 'Last-frontier Minerals Scramble',
     The Canberra Times, 16 February 1983, p.31;
     'Mountain of Iron in the Antarctic', *Canberra Times*,
     20 December 1976, p.4.

10   'Move to Prevent Antarctic Exploitation: Plan
     for Polar World Park', *The Canberra Times*,
     13 September 1983, p.7.

11   Greenpeace Australia was established in 1977 as
     part of Greenpeace International. See Guide to
     the Records of Greenpeace Australia (Antarctic
     Campaign), National Library of Australia, MS 9432.

12   'Greenpeace History in the Antarctic', documents.ats.
     aq/ATCM25/ip/ATCM25_ip101_e.pdf.

13   Sam Blay and Ben M. Tsamenyi, 'Australia and the
     Convention for the Regulation of Antarctic Mineral
     Resource Activities (CRAMRA)', *Faculty of Law,
     Humanities and the Arts – Papers*, no.234, 1990,
     p.198; Griffiths, *Slicing the Silence*, pp.286–287.

14   *The Sydney Morning Herald*, 22 March 1989, cited
     in Blay and Tsamenyi, 'Australia and the Convention
     for the Regulation of Antarctic Mineral Resource
     Activities (CRAMRA)', p.198.

15   Secretariat of the Antarctic Treaty, *The Protocol on
     Environmental Protection to the Antarctic Treaty*,
     ats.aq/e/protocol.html.

16   For a detailed study of Australia's role in the
     Antarctic Treaty System, see Andrew Jackson,
     *Who Saved Antarctica? The Heroic Era of Antarctic
     Diplomacy*, Switzerland: Palgrave Macmillan, 2021;
     Andrew Jackson and Lorne Kriwoken, 'The Protocol
     in Action, 1992–2010', in Marcus Haward and Tom
     Griffiths (eds), *Australia and the Antarctic Treaty
     System: 50 Years of Influence*, Sydney: UNSW Press,
     2011, pp.300–319.

17   Maj de Poorter, 'Voyage to Antarctica, 9 February
     1987', in John May, *The Greenpeace Book of
     Antarctica: A New View of the Seventh Continent*,
     Frenchs Forest: Child & Associates, 1989, p.163.

18   Australian Antarctic Division, 'Waste Management',
     antarctica.gov.au/about-antarctica/environment/
     pollution-and-waste/waste.

19   Richard Penney, cited in Clark and Wishart,
     66 South, p.94.

20   Murray-Smith, *Sitting on Penguins*, p.85.

21   'Casey: the Daintree of Antarctica', antarctica.gov.au/
     about-antarctica/wildlife/plants/casey-the-daintree-
     of-antarctica.

22   James Fenton and Ronald Smith, 'Distribution,
     Composition and General Characteristics of the
     Moss Banks of the Maritime Antarctic', *British
     Antarctic Survey Bulletin*, vol.51, 1982, pp.215–236;

23   Joan Russell, cited in Clark and Wishart, *66 South*,
     p.94.

24   Tas van Ommen, interview with author, Hobart,
     27 March 2021.

25   Mawson, *Home of the Blizzard*, vol.1, p.255.

26   J.W. Gregory, *The Climate of Australasia in Reference
     to its Control By the Southern Ocean*, Melbourne:
     Whitcombe and Tombs Limited, 1904, p.30.

27   The term Little Ice Age, coined in 1939, refers to the
     period between 1300 and 1850 CE when mountain
     glaciers expanded to their greatest extent. It followed
     the Medieval Warming Period (900 to 1300 CE)
     and preceded the present period of warming from
     the late nineteenth century. See John P. Rafferty and
     Stephen T. Jackson, 'Little Ice Age', *Encyclopedia
     Britannica*, 7 April 2011, britannica.com/science/
     Little-Ice-Age.

28   Initially called the Wilkes ice cap, the Law Dome
     was renamed in 1970 in honour of Phillip Law. Ice
     in the Law Dome is around 90,000 years old, but the
     bulk of it is only 4,000 to 10,000 years old because
     the oldest ice—representing the earliest 70,000
     years—is compressed at the bottom. See 'What Ice
     Cores from Law Dome Can Tell Us about Past and
     Current Climates', interview with Tas van Ommen
     by Joseph Cheek, *SciencePoles*, 12 August 2011,
     sciencepoles.org/interview/what-ice-cores-from-law-
     dome-can-tell-us-about-past-and-current-climates;
     see also Australian Antarctic Division, 'The Power of
     Three—Rainfall, Ice Cores and Climate Models—in
     Australian Water Management', 24 August 2020,
     antarctica.gov.au/news/2020/the-power-of-three-
     rainfall-ice-cores-and-climate-models-in-australian-
     water-management. William F. Budd, 'The Antarctic
     Ice Shelf', *Australian Antarctic Science: The First Fifty
     Years of ANARE*, Kingston: Australian Antarctic
     Division, 2003, pp.320–321.

29   Tas van Ommen, 'Antarctic Ice Cores Shed Light on
     Western Australian Drought', *Australian Antarctic
     Magazine*, issue 18, 2010; Australian Antarctic
     Division, 'Antarctic Ice Cores Tell 1000-year
     Australian Drought Story', 16 December 2014; Cheek,
     'What Ice Cores from Law Dome Can Tell Us'.

30   van Ommen, interview with author, Hobart,
     27 March 2021.

31   van Ommen, interview with author, Hobart,
     27 March 2021; Tas van Ommen et al., *Position
     Analysis: Antarctic Ice Cores and Climate Change*,
     Hobart: Antarctic Climate and Ecosystems
     Cooperative Research Centre, 2015, p.10.

32 The Antarctic ice sheet is regarded as the richest
    source of information about Earth's climate history.
    van Ommen et al., *Position Analysis*, p.5; Cheek,
    'What Ice Cores from Law Dome Can Tell Us'.
33 The 800,000-year-old ice core was drilled from ice
    3,270 metres thick. See, for example, Ed Brook,
    'Windows On the Greenhouse', *Nature*, vol.453, 2008,
    pp.291–292.
34 van Ommen, interview with author, Hobart,
    27 March 2021.
35 van Ommen, interview with author, Hobart,
    27 March 2021.
36 van Ommen, interview with David Sparkes,
    'Antarctic Odyssey: Epic Journey in Search of the
    "Million-year Ice Core"', *ABC Radio*, 28 June 2019.
    Other nations are also pursuing drilling programs
    in search of the oldest, deepest ice cores.
37 Graham Readfearn, '"Colder and Deeper": Scientists
    Close in on Spot to Drill Antarctic Ice Core 1.5m
    Years Old', *The Guardian*, 13 February 2021,
    theguardian.com/environment/2021/feb/13/colder-
    and-deeper-scientists-close-in-on-spot-to-drill-
    antarctic-ice-core-15m-years-old.
38 van Ommen, interview with author, Hobart,
    27 March 2021.
39 National Snow & Ice Data Center, nsidc.org/
    cryosphere/quickfacts/iceshelves.html.
40 Australian Antarctic Division, 'Ocean Waves
    Following Sea Ice Loss Trigger Antarctic Ice Shelf
    Collapse', 14 June 2018, antarctica.gov.au/news/2018/
    ocean-waves-following-sea-ice-loss-trigger-antarctic-
    ice-shelf-collapse.
41 In 2014, the estimated population of emperor
    penguins, which live exclusively in Antarctica, was
    238,000 breeding pairs. P.T. Fretwell et al., 'Emperor
    Penguins Breeding on Iceshelves', *PLOS ONE*, vol.9,
    no.1, 8 January 2014, e85285.
42 Patrick G. Quilty, 'Influences on the Future
    Directions of Australian Antarctic Research', in
    Marchant et al. (eds), *Australian Antarctic Science*,
    pp.581–582.
43 Alessandro Silvano, et al., 'Freshening by Glacial
    Meltwater Enhances Melting of Ice Shelves and
    Reduces Formation of Antarctic Bottom Water',
    Science Advances, vol.4, no.4, 18 April 2018,
    science.org/doi/10.1126/sciadv.aap9467; also see
    Helen Phillips et al., 'Explainer: How the Antarctic
    Circumpolar Current Helps Keep Antarctica
    Frozen', *The Conversation*, 16 November 2018,
    theconversation.com/explainer-how-the-antarctic-
    circumpolar-current-helps-keep-antarctica-frozen;
    Katherine Woodthorpe and Tony Press, 'How Can
    Antarctic Science Help Us Understand the Australian
    Climate?', Australian Academy of Technology and
    Engineering, 6 October 2020, atse.org.au/news-
    and-events/article/how-antarctic-science-helps-us-
    understand-australian-climate.
44 Griffiths, *Slicing the Silence*, pp.33 and 69;
    '"Antarctic Air" to Buffet Australia's South-east
    as Weather Warning Issued to Bushwalkers', *The
    Guardian*, 3 August 2020, theguardian.com/
    australia-news/2020/aug/03/antarctic-air-to-buffet-
    australias-south-east-as-weather-warning-issued-
    to-bushwalkers; '"Unusual" Antarctic Polar Vortex
    Could Mean Cold Wet Aussie Summer', *9News*,
    12 November 2020, 9news.com.au/wild-weather/
    polar-vortex-above-antarctica-could-mean-wetter-
    colder-summer-for-australia.

Select bibliography and suggestions for further reading or viewing

Abercromby, Ralph (ed.), *Three Essays on Australian Weather*, Sydney: F.W. White, 1896.

Antarctic Exploration: Baron Nordenskiöld's Proposal for further Antarctic exploration. With notes on the same by Alfred J. Taylor, Hobart, 1890.

Antonello, Alessandro, *The Greening of Antarctica: Assembling an International Environment*, New York: Oxford University Press, 2019.

Ayres, Philip, *Mawson: A Life*, Melbourne: Miegunyah Press, 1999.

Barnes, David, 'Antarctic Seabed Carbon Capture Change', British Antarctic Survey website.

Barr, Susan, and Lüdecke, Cornelia (eds), *The History of the International Polar Years (IPYs)*, Berlin: Springer-Verlag, 2010.

Bax, Narissa, et al., 'Perspective: Increasing Blue Carbon around Antarctica is an Ecosystem Service of Considerable Societal and Economic Value Worth Protecting', *Global Change Biology*, vol.27, no.1, 2021.

Bernacchi, Louis Charles, *To the South Polar Regions: Expedition of 1898–1900*, London: Hurst and Blackett, 1901.

Béchervaise, John, *Blizzard and Fire: A Year at Mawson, Antarctica*, Sydney: Angus and Robertson, 1963.

Béchervaise, John, *Australia and Antarctica*, Lane Cove: Nelson Doubleday Pty Ltd, 1967.

Bindoff, Nathan, et al., *Position Analysis: Climate Change and the Southern Ocean*, Hobart: Antarctic Climate and Ecosystems Cooperative Research Centre, 2011.

Black, Harry, 'Jumbo Jet to Antarctica', *Aurora: ANARE Club Journal*, Midwinter 1977.

Blackadder, Jesse, 'A Language for Antarctica: Summer at Mawson', *Kill Your Darlings*, 11 June 2019 (online).

Bowden, Tim, *The Silence Calling: Australians in Antarctica 1947–97*, The ANARE Jubilee History, Sydney: Allen & Unwin, 1997.

Branagan, David, *T.W. Edgeworth David: A Life*, Canberra: National Library of Australia, 2004.

Broecker, Wallace S., *The Great Ocean Conveyor: Discovering the Trigger for Abrupt Climate Change*, Princeton: Princeton University Press, 2010.

Budd, Graeme M., 'Australian Exploration of Heard Island, 1947–1971', *Polar Record*, vol.43, no.225, 2007.

Bull, H.J., *The Cruise of the 'Antarctic': To the South Polar Regions*, London and New York: Edward Arnold, 1896; 1984 edition (originally published in Norwegian as *Sydover: Ekspeditionen til Sydishavet i 1893–1895*).

Burns, Robin, *Just Tell Them I Survived!: Women in Antarctica*, Sydney: Allen & Unwin, 2001.

Cheek, Joseph, 'What Ice Cores from Law Dome Can Tell Us about Past and Current Climates', *SciencePoles*, 12 August 2011.

Cherry-Garrard, Apsley, *The Worst Journey in the World*, London: Vintage, 2010 (1922).

Chipman, Elizabeth, *Women on the Ice: A History of Women in the Far South*, Carlton: Melbourne University Press, 1986.

Clark, Linda, and Wishart, Elspeth, *66 South: Tales from an Antarctic Station*, Launceston: Queen Victoria Museum and Art Gallery, 1993.

Coleridge, Samuel Taylor, 'The Rime of the Ancient Mariner', in William Wordsworth and Samuel Taylor Coleridge, *Lyrical Ballads, with a Few Other Poems*, vol.1, London: J & A Arch, 1798.

Collis, Christy, 'Mawson's Hut: Emptying Post-colonial Antarctica', *Journal of Australian Studies*, vol.23, no.63, 1999.

Collis, C., 'The Proclamation Island Moment: Making Antarctica Australian', *Law Text Culture*, vol.8, article 3, 2004 (online).

Commonwealth of Australia, *Antarctic Journey: Three Artists in Antarctica*, Canberra: Australian Government Publishing Service, 1988.

Cook, James, *A Voyage Towards the South Pole, and Round the World, Performed in His Majesty's Ships the Resolution and Adventure, in the Years 1772, 1773, 1774, and 1775*, vol.1, book 1, June 1772, London: W. Strahan & T. Cadell, London, 1777.

Crane, Ralph, et al., *Imagining Antarctica: Cultural Perspectives on the Southern Continent*, Hobart: Quintus Publishing, 2011.

Crawford, Janet, *That First Antarctic Winter: The Story of the Southern Cross Expedition of 1898–1900 As Told in the Diaries of Louis Charles Bernacchi*, Christchurch: South Latitude Research Limited/Peter J Skellerup, 1998.

CSIRO Australia, 'Southern Ocean Winds Open Window to the Deep Sea', *Science Daily*, 17 March 2010 (online).

Cumpston, John Stanley, *Macquarie Island*, Canberra: Australian Antarctic Division, Department of External Affairs (Australia), 1968.

Dakin, W.J., *Whalemen Adventurers*, Sydney: Angus and Robertson, 1938.

David, T.W. Edgeworth, *The First Journey to the South Magnetic Pole: Part of the Story of the British Antarctic Expedition 1907–1909*, Philadelphia: The Washington Square Press, 1909.

Davis, J.K., *High Latitude*, Parkville: Melbourne University Press, 1962.

Dixon, Robert, *Photography, Early Cinema and Colonial Modernity: Frank Hurley's Synchronized Lecture Entertainments*, Anthem Australian Humanities Research Series, London: Anthem Press, 2011.

Dodds, Klaus, et al. (eds), *Handbook on the Politics of Antarctica*, Cheltenham, UK: Edward Elgar Publishing Inc., 2017.

Fenton, James, and Smith, Ronald, 'Distribution, Composition and General Characteristics of the Moss Banks of the Maritime Antarctic', *British Antarctic Survey Bulletin*, vol.51, 1982.

Fogg, Gordon Elliott, *A History of Antarctic Science*, Cambridge: Cambridge University Press, 1992.

Fretwell, P.T., et al., 'Emperor Penguins Breeding on Iceshelves', *PLOS ONE*, vol.9, no.1, 8 January 2014.

Gantevoort, M., et al., 'Reconstructing Star Knowledge of Aboriginal Tasmanians', *Journal of Astronomical History and Heritage*, vol.19, no.3, 2016.

Gillham, Mary E., *Sub-Antarctic Sanctuary: Summertime on Macquarie Island*, Wellington: A.H. and A.W. Reed, 1967.

Gogarty, Brendan, et al., 'Protecting Antarctic Blue Carbon: As Marine Ice Retreats Can the Law Fill the Gap?', *Climate Policy*, vol.20, no.2, 2020.

Greenwood, Gordon, and Grimshaw, Charles (eds), *Documents on Australian International Affairs 1901–1918*, West Melbourne: Thomas Nelson (Australia) with London: Australian Institute of International Affairs and the Royal Institute of International Affairs, 1977.

Gregory, J.W., *The Climate of Australasia in Reference to Its Control by the Southern Ocean*, Melbourne: Whitcombe and Tombs Limited, 1904.

Griffiths, Tom, 'Australia's Culture of the Ice—Ninth Phillip Law Lecture', *The Phillip Law Lectures*, Department of Economic Development, Tourism and the Arts, Hobart, 31 July 2011.

Griffiths, Tom, 'Debunking Mawson', *Inside Story*, 3 December 2013 (online).

Griffiths, Tom, *Slicing the Silence: Voyaging to Antarctica*, Sydney: NewSouth, 2007.

Griffiths, Tom, 'The Breath of Antarctica', in *Tasmanian Historical Studies*, vol.11, 2006 (online).

Hains, Brigid, *The Ice and the Inland: Mawson, Flynn, and the Myth of the Frontier*, Carlton: Melbourne University Press, 2002.

Hamacher, Duane W., 'Aurorae in Australian Aboriginal Traditions', *Journal of Astronomical History and Heritage*, vol.17, no.1, 2014 (online)

Harris, Nisha, 'Antarctic Bottom Water Disappearing', *Australian Antarctic Magazine*, no.23, December 2012.

Haward, Marcus, and Griffiths, Tom (eds), *Australia and the Antarctic Treaty System: 50 Years of Influence*, Sydney: UNSW Press, 2011.

Hayes, J.G., *Antarctica: A Treatise on the Southern Continent*, London: Richards Press, 1928.

Hince, Bernadette, *The Antarctic Dictionary: A Complete Guide to Antarctic English*, Canberra: CSIRO Publishing, 2000.

Hince, Bernadette (ed.), *Still No Mawson: Frank Stillwell's Antarctic Diaries 1911–13*, Canberra: Australian Academy of Science, 2012.

Hince, Bernadette, et al. (eds), *Antarctica: Music, Sounds and Cultural Connections*, Canberra: ANU Press, 2015.

Hunt, H.A., et al., *The Climate and Weather of Australia*, Commonwealth Bureau of Meteorology, 1913.

Hurley, Frank, *Shackleton's Argonauts: A Saga of the Antarctic Ice-Packs*, Sydney: Angus and Robertson, 1948.

International Council of Scientific Unions, World Meteorological Organization, Intergovernmental Oceanographic Commission and Scientific Committee on Oceanic Research, *Scientific Plan for the World Ocean Circulation Experiment*, World Climate Research Programme, WCRP Publications Series, no.6, WMO/TD no.122, July 1986.

Jackson, Andrew, *Who Saved Antarctica? The Heroic Era of Antarctic Diplomacy*, Switzerland: Palgrave Macmillan, 2021.

Jarvis, Tim, *Mawson: Life and Death in Antarctica*, Carlton: Miegunyah–Melbourne University Press, 2008.

Joerg, W.L.G., (ed.), *Problems of Polar Research*, American Geographical Society Special Publication no.7, New York, 1928.

Kawaja, Marie, 'Australia in Antarctica: Realising an Ambition', *The Polar Journal*, vol.3, no.1, 2013.

Keneally, Thomas, 'Captain Scott's Biscuit: An Antarctic Pilgrimage', *Granta*, no.83, 2003 (online).

Kriwoken, Lorne K., and Williamson, John W., 'Hobart, Tasmania: Antarctic and Southern Ocean Connections', *Polar Record*, vol.29, no.169, April 1993.

Laseron, Charles Francis, *South with Mawson: Reminiscences of the Australasian Antarctic Expedition, 1911–1914*, London and Sydney: George G. Harrap and Company Ltd in association with Australasian Publishing Co Pty Ltd, 1947.

Law, Phillip, *Antarctic Odyssey*, Melbourne: Heinneman, 1983.

Law, Phillip, *Australia and the Antarctic*, The John Murtagh Macrossan Memorial Lectures, 1960, St Lucia: University of Queensland Press, 1962.

Law, Phillip, 'Australian Coastal Exploration in Antarctica', *The Geographical Journal*, vol.124, no.2, 1958.

Law, Phillip, 'Australian National Antarctic Research Expedition, 1955', *The Geographical Journal*, vol.122, no.1, 1956.

Law, Phillip, 'The IGY in Antarctica', *The Australian Journal of Science*, vol.21, no.9, 1959.

Lawrence, Susan, and Staniforth, Mark (eds), *The Archaeology of Whaling in Southern Australia and New Zealand*, Gundaroo: Brolga Press for Australasian Society for Historical Archaeology and Australian Institute for Maritime Archaeology, 1998.

Leane, Elizabeth, 'A Place of Ideals in Conflict: Images of Antarctica in Australian Literature', in C.A. Cranston and Robert Zeller (eds), *The Littoral Zone: Australian Contexts and their Writers*, Amsterdam: Rodopi, 2007.

Leane, Elizabeth (comp.), 'Representations of Antarctica', a bibliography of literature and film (online).

Leane, Elizabeth, *South Pole: Nature and Culture*, London: Reaktion Books Ltd, 2016.

Leane, Elizabeth, and Tiffin, Helen, 'Dogs, Meat and Douglas Mawson', *Australian Humanities Review*, issue 51, November 2011 (online).

Leane, Elizabeth, et al., 'Beyond the Heroic Stereotype: Sidney Jeffryes and the Mythologising of Australian Antarctic History', *Australian Humanities Review*, issue 64, May 2019 (online).

McCann, Joy, *Wild Sea: A History of the Southern Ocean*, Sydney: NewSouth, 2018.

McGregor, Alasdair, *Frank Hurley: A Photographer's Life*, Camberwell: Viking, 2004.

McIntyre, Don (with Peter Meredith), *Two Below Zero: A Year Alone in Antarctica*, Terrey Hills: Australian Geographic, 1996.

McKay, Helen F. (ed.), *Gadi Mirrabooka: Australian Aboriginal Tales from the Dreaming*, Eaglewood: Libraries Unlimited, 2001.

Macquarie Island Nature Reserve and World Heritage Area Management Plan 2006, Hobart: Parks and Wildlife Service, 2006.

Madigan, Cecil T., *Tabulated and Reduced Records of the Cape Denison Station, Adélie Land, Australasian Antarctic Expedition 1911–14*, Scientific Reports, series B, vol.IV, Meteorology, Sydney: Alfred J Kent, Government Printer, 1929.

Markham, Clements R., 'The Need for an Antarctic Expedition', in *The Nineteenth Century: A Monthly Review*, March 1877–December 1900, vol.38, issue 224, London, October 1895.

Marchant, Harvey J., et al. (eds), *Australian Antarctic Science: The First 50 Years of ANARE*, Kingston: Australian Antarctic Division, 2002.

Mawson, Douglas, *AAE Scientific Reports A, vol.1, Narrative and Cartography*, Sydney: Government Printer, 1943.

Mawson, Douglas, *Geographical Narrative and Cartography*, Australasian Antarctic Expedition 1911–14, Scientific Reports, series A, vol.1, part 1, Sydney: Government Printer, 1942.

Mawson, Sir Douglas, *Macquarie Island: A Sanctuary for Australasian Sub Antarctic Fauna: a lecture*, delivered before the Royal Geographical Society of Australasia (South Australian Branch) Incorporated, Adelaide, 12 September 1919.

Mawson, Douglas, *Macquarie Island: Its Geography and Geology*, Australasian Antarctic Expedition 1911–14, Scientific Reports, series A, vol.5, Sydney: Government Printer, 1943.

Mawson, Sir Douglas, *The Home of the Blizzard*, London: William Heinemann, 1915. For online access see 'Mawson's book: The Home of the Blizzard', Australian Antarctic Division, antarctica.gov.au/about-antarctica/history/exploration-and-expeditions/australasian-antarctic-expedition/mawsons-book.

May, John, *The Greenpeace Book of Antarctica: A New View of the Seventh Continent*, Frenchs Forest: Child & Associates, 1989.

Mosley, Geoff, *Saving the Antarctic Wilderness: The Pivotal Role in its Complete Protection*, Canterbury: Envirobook, 2009.

Moyes, J., *Exploring the Antarctic with Mawson and the Men of the 1911–1914 Expedition*. West Gosford: John L. Moyes, 1996.

Murray-Smith, Stephen, *Sitting on Penguins: People and Politics in Australian Antarctica*, Surry Hills: Hutchinson Australia, 1988.

Nash, Michael, *The Bay Whalers: Tasmania's Shore-based Whaling Industry*, Woden: Navarine, 2003.

Parer-Cook, Elizabeth (with Neil Roberts), *Mawson 1972: Reflections 40 Years On*, Brighton: Zaurora Books, 2012.

Patterson, Diana, *The Ice Beneath My Feet: My Year in Antarctica*, Sydney: ABC Books, 2010.

Philpott, Carolyn, and Leane, Elizabeth, 'Making Music on the March: Sledging Songs of the "Heroic Age" of Antarctic Exploration', *Polar Record*, vol.52, no. 6, 2016.

Press, Tony, 'In Deep: Australian Research in the Southern Ocean', *Australian Antarctic Magazine*, no.4, Spring 2002.

Price, A. Grenfell, *The Winning of Australian Antarctica: Mawson's BANZARE Voyages, 1929–31, based on the Mawson papers*, published by Mawson Institute for Antarctic Research, University of Adelaide, Sydney: Angus and Robertson, 1962.

Ptolemy, Claudius, *Geographia*, trans. Giacomo d'Angelo da Scarperia, Bologna: Dominicus de Lapis, 1477 (c. 150 CE).

Pyne, Stephen J., *The Ice*, London: Weidenfeld & Nicolson, 1986.

Pyper, Wendy, 'Southern Ocean Marine Life in Focus', *Australian Antarctic Magazine*, issue 18, 2010.

Quilty, P., 'Introducing Antarctica', *Issues: 'Why Does the World need Antarctica?'*, Melbourne: Australian Council for Education Research, March 1996.

Rafferty, John P., and Jackson, Stephen T., 'Little Ice Age', *Encyclopedia Britannica*, 7 April 2011 (online).

Readfearn, Graham, '"Colder and Deeper": Scientists Close

in on Spot to Drill Antarctic Ice Core 1.5m Years Old', *The Guardian*, 13 February 2021 (online).

Rintoul, Stephen, et al., 'The Southern Ocean in a Changing Climate', Project R1.1, Hobart: Antarctic Climate and Ecosystems Cooperative Research Centre.

Russell, Lynette, *Roving Mariners: Australian Aboriginal Whalers and Sealers in the Southern Oceans, 1790–1870*, Albany: State University of New York Press, 2012.

Sarris, Aspa, 'Antarctic Station Life: The First 15 Years of Mixed Expeditions to the Antarctic', *Acta Astronautica*, vol.131, 2017.

Scambos, T.A., et al., 'Ultralow Surface Temperatures in East Antarctica from Satellite Thermal Infrared Mapping: The Coldest Places on Earth', *Geophysical Research Letters*, vol.45, no.12, 2018.

Schillat, Monika, 'Images of Antarctica as Transmitted by Literature' in *Tourism in Antarctica*, Springer Briefs in Geography, New York: Springer, 2016.

Scholes, Arthur, *Fourteen Men: The Story of the Australian Antarctic Expedition to Heard Island*, Melbourne: F.W. Cheshire, 1949.

'Secret Instructions for Lieutenant James Cook Appointed to Command His Majesty's Bark the *Endeavour*', 30 July 1768, Canberra: National Library of Australia, MS 2.

Shackleton, Sir Ernest Henry, *The Heart of the Antarctic; Being the Story of the British Antarctic Expedition 1907–1909, by E.H. Shackleton, CVO, with an introduction by Hugh Robert Mill, DSc, an account of the first journey to the south magnetic pole by Professor T.W. Edgeworth*, London: W. Heinemann, 1909.

Sherratt, Tim, et al. (eds), *A Change in the Weather: Climate and Culture in Australia*, Canberra: National Museum of Australia Press, 2005.

Simpson, Margaret, 'Extreme South: James Castrission and Justin Jones' Antarctic adventure', Sydney: Museum of Applied Arts & Sciences, 2 August 2012 (online).

'Sinking a Small Fortune: Joseph Hatch and the Oiling Industry', *The Sealers' Shanty: A Journal of Sealers' Stories*, vol.9: 1889–1919, 13 April 2017 (online).

Sixth International Geographical Congress, London, 1895, in *Journal of the American Geographical Society of New York*, vol.27, no.3, 1895.

Souter, Gavin, *Acts of Parliament: A Narrative History of the Senate and House of Representatives*, Carlton: Melbourne University Press, 1988.

Spufford, Francis, 'The Uses of Antarctica: Roles for the Southern Continent in Twentieth-century Culture', in Ralph Crane et al. (eds) *Imagining Antarctica: Cultural Perspectives on the Southern Continent*, Hobart: Quintus, 2011.

Stevenson, R.L., *The Letters of Robert Louis Stevenson to His Family and Friends*, selected and edited with notes and introduction by Sidney Colvin, London: Methuen and Co, 1899.

Swan, R.A., *Australia in the Antarctic: Interest, Activity and Endeavour*, Carlton: Melbourne University Press, 1961.

Taylor, Griffith, *Australian Meteorology: A Text-Book Including Sections on Aviation and Climatology*, Oxford: Clarendon Press, 1920.

Thomson, Sir Charles Wyville, *The Voyage of the 'Challenger'. The Atlantic: A Preliminary Account of the General Results of the Exploring Voyage of H.M.S. Challenger During the Year 1873 and the Early Part of the Year 1876*, vol.2, Cambridge: Cambridge University Press, 2014 (1877).

Thomson, Robert, *The Coldest Place on Earth*, Wellington: A.H. & A.W. Reed, 1969.

Tizard, Thomas Henry, et al., *Narrative of the Cruise of H.M.S. Challenger, with a General Account of the Scientific Results of the Expedition*, vol.1 of Great Britain, *Challenger* Office, Report on the Scientific Results of the Voyage of H.M.S. *Challenger* During the Years 1873–76 under the Command of Captain George S. Nares, R.N., F.R.S. and the Late Captain Frank Tourle Thomson, R.N., Edinburgh: Neill, 1885.

van Ommen, Tas, et al., *Position Analysis: Antarctic Ice Cores and Climate Change*, Hobart: Antarctic Climate and Ecosystems Cooperative Research Centre, 2015.

Wheeler, Barbara, and Young, Linda, 'Antarctica in museums: the Mawson collections in Australia', *Polar Record*, vol.36, no.198, 2000.

Wiseman, Rowena, et al. (eds), *Sidney Nolan: Antarctic Journey*, Melbourne: Mornington Peninsula Regional Gallery, 2006.

Zillman, John, *A Hundred Years of Science and Service: Australian Meteorology Through the Twentieth Century*, Australian Bureau of Statistics, 1301.0 – *Yearbook Australia*, 2001.

Newspapers

The Advertiser (Adelaide), 8 September 1914

The Age (Melbourne), 9 February 1954

The Canberra Times, 16 February 1983; 13 September 1983; 20 December 1976

Commercial Journal and Advertiser, 18 April 1840

The Courier (Hobart, Tas), 4 May 1841

Evening Journal (Adelaide), 27 July 1910

Evening Star (Dunedin), no. 9807, 21 September 1895

Hobart Town Advertiser, 28 February 1840

Independent (UK), 2 September 1998

The Mercury (Hobart), 27 December 1933

The Observer (Adelaide), 7 March 1914

The Register (Adelaide), 28 April 1910

The Sydney Morning Herald, 25 April 1891; 22 March 1989

The Sydney Gazette and New South Wales Advertiser, 15 April 1820; 13 December 1822; 2 June 1831

Tasmanian News (Hobart), 13 June 1895

Interviews by the author

Denise Allen, 15 November 2019
Dr Narissa Dax, 14 November 2019
Dr Jaimie Cleeland, 17 September 2019
Keith Gooley, 16 November 2019
Dr Tas van Ommen, 27 March 2021
Dr Barbara Wienecke, 30 September 2019

Manuscripts

National Library of Australia (NLA)

Records relating to the Antarctic region can be found in many of the Library's manuscript collections. The following are the principal collections documenting expeditions in Antarctica:

Béchervaise, John, leader of three ANARE expeditions to Heard Island and Mawson (1953–1960), papers, NLA MS 7972.

Brown, D.A. (Duncan Alexander), a member of the ANARE expeditions to Macquarie Island (1956), Mawson Station (1958) and Davis Station (1961), papers, NLA MS 9452.

Chipman, Elizabeth, papers documenting her work in the Antarctic Division, her three visits to Macquarie Island (1967–1975), and her research on Australians in the Antarctic, NLA MS 9635.

Davis, Captain John King, letters and cuttings concerning his book *High Latitude*, 1962–1963, NLA MS 9234.

Elliott, Fred, a weather observer on ANARE expeditions to Heard Island and Mawson, 1953–1958, diaries and scrapbooks, NLA MS 9442.

Gibbney, Leslie, a biologist on the 1950 and 1952 ANARE expeditions to Heard Island, diaries, NLA MS 9392.

Guide to the Records of Greenpeace Australia (Antarctic Campaign), 1981–1994, NLA MS 9432.

Hodgson, Thomas Vere, member of the 1901–4 British National Antarctic Expedition, papers, NLA MS 223.

Hunter, John, chief biologist on the Australasian Antarctic Expedition, 1911–1914, diaries and papers, NLA MS 2806.

Hurley, Frank, the photographer on the Antarctic expeditions led by Douglas Mawson (1911–1914 and 1929–1931) and Ernest Shackleton (1914–1917), papers, NLA MS 883.

Jones, Sydney Evan, a medical officer on the Australasian Antarctic Expedition, 1911–1914, a diary and papers, NLA MS 9273.

Law, Phillip Garth, senior scientific officer (1947–1948) and leader of 23 ANARE expeditions to the Antarctic and subantarctic (1949–1966), papers, NLA MS 9458.

Mawson, Sir Douglas, papers mostly relating to Macquarie

Island. They include diaries and logs (1912–1913), a report on the island's soils (1915), and logs kept on the *Discovery* (1930–1931), NLA MS 6650.

Scholes, Arthur, diaries kept on the ANARE expedition to Heard Island, 1947–1951, NLA MS 203.

'Secret Instructions for Lieutenant James Cook Appointed to Command His Majesty's Bark the Endeavour', 30 July 1768, NLA MS 2.

Summers, Robert, a medical officer on the ANARE expedition to Mawson in 1954, diary and scrapbooks, NLA MS 9165.

Taylor, Thomas Griffith, the senior geologist on the British Antarctic Expedition in 1910–1913, papers, NLA MS 1003.

Whiting, John C., a pilot who undertook survey flights in Antarctica in 1965, papers and photographs, NLA MS 6595.

Mitchell Library, State Library of New South Wales

Harrisson, Charles Turnbull, diary, 2 December 1911–31 December 1912, Mitchell Library, ML MSS 386.

Hurley, Frank, sledging diary, 10 November 1912–10 January 1913, kept while a member of the Australasian Antarctic Expedition, 1911–1914, together with an edited typescript transcript, Mitchell Library, ML MSS 389/1.

Laseron, Charles Francis, diaries, 21 November 1911–24 February 1913, kept while a member of the Australasian Antarctic Expedition, 1911–1914, together with related papers, 1911–ca. 12 October 1912, Mitchell Library, ML MSS 385.

Moyes, Morton Henry, diary, 2 December 1911–23 February 1913, Mitchell Library, ML MSS 388.

Wild, Frank, papers, c. 1921–1937, vol.1, memoirs, 1937?, Mitchell Library, ML MSS 2198.

Other manuscripts

Erceg, Diane, 'Explorers of a Different Kind: A History of Antarctic Tourism 1966–2016', PhD thesis, Australian National University, 2017.

Hince, Bernadette, 'The Teeth of the Wind: An Environmental History of Subantarctic Islands', PhD thesis, Australian National University, 2005.

Kawaja, Marie, 'The Politics and Diplomacy of the Australian Antarctic, 1901–1945', PhD thesis, Australian National University, May 2010.

Worsley, Frank, collection, Scott Polar Research Institute Archives, University of Cambridge, GB15 Frank Worsley.

National, state and territory collections

Collections relating to Australia's Antarctic history are distributed across a wide range of institutions. Relevant material is also held in Australian private collections, as well as in overseas collections, notably in New Zealand and the United Kingdom. The following presents a selection of publicly accessible collections in Australia.

National Library of Australia

The National Library of Australia holds significant collections relating to Australians and Antarctica, including the Australian Antarctic Division oral history collection and Australian Antarctic culture oral history project; Antarctic books for children; manuscripts, photographs and maps donated by Australians associated with Antarctica; publications and reports; diaries and papers written by some early Australians involved in Antarctic exploration and science; records of the Australian National Research Council, including minutes and papers of its Antarctic Committee (1927–1935) and papers on an Antarctic expedition (1947); an extensive map collection relating to Australian voyages and flights in the Antarctic; and collections of Antarctic photographs, including images taken by Frank Hurley and Douglas Mawson.

Australian Antarctic Division (AAD)

A large collection of online resources including:

- scientific databases, historical information about ANARE, and the Division's activities and resources
- a bimonthly newsletter, weekly station news updates and regular news items
- live webcams at the three continental stations (Mawson, Casey and Davis), on the subantarctic station at Macquarie Island, and on board *Nuyina*
- *Home of the Blizzard: The Australasian Antarctic Expedition*—a valuable website with images, maps and videos relating to the AAE in 1911–1914
- Australian Antarctic Arts Fellowship, Australian Antarctic Division website, antarctica.gov.au
- *Nuyina* website, nuyina.antarctica.gov.au
- Australian Antarctic Program social media: @AusAntarctic.

ANARE Club

The ANARE Club maintains a national website, including news for members and access to Antarctic websites and weather webcams.

National Archives of Australia (NAA)

The NAA holds records relating to Australian Antarctic exploration and research.

Australian Academy of Science

The Academy of Science provides online access to interviews with selected scientists who have worked in the Antarctic and subantarctic regions.

Screen Australia

Screen Australia (National Film and Sound Archive of Australia) provides online access to historical film footage, from Mawson's first expedition and the establishment of Australia's Antarctic stations in the 1950s to modern documentaries and sound recordings relating to Antarctica. These include Frank Hurley's films: *The Home of the Blizzard: Life in the Antarctic* (The Official Film of the Mawson Antarctic Expedition (c.1916); *Endurance* (1917); and *Siege of the South* (1931).

Australian Dictionary of Biography

The Australian Dictionary of Biography (ADB) provides online access to entries about the lives of notable Australians associated with Antarctica.

Geoscience Australia

Geoscience Australia documents the history of the organisation's mapping and exploration work in Antarctica.

Other repositories

Australian State and Territory libraries, museums and archives hold important collections relating to Australians in Antarctica, including the State Library of New South Wales, Museums Victoria, Australian Museum, National Museum of Australia, State Library of South Australia, Tasmanian Museum and Art Gallery, Museum of Applied Arts and Sciences (Powerhouse Museum), Queensland Museum, Western Australian Museum, and Royal Botanic Garden, Sydney. In particular, the SA Museum holds the Australian Polar Collections relating to the South Australian explorers Sir Douglas Mawson, Sir George Hubert Wilkins and John Riddoch Rymill. Mawson's collection contains his personal artefacts, including the balaclava he wore in Frank Hurley's portrait and as depicted on the first Australian $100 note. The University of Tasmania manages the Australian Collection of Antarctic Microorganisms. Australia Post holds collections of stamps relating to Australians and Antarctica. The University of New England holds the Griffith Taylor Collection. The University of Adelaide holds the Mawson Antarctic Collection.

List of Illustrations

Where indicated, images are reproduced under: CC BY 2.0: creativecommons.org/licenses/by/2.0/, CC BY-SA 2.0: creativecommons.org/licenses/by-sa/2.0/; CC BY 3.0: creativecommons.org/licenses/by/3.0/; CC BY-SA 3.0: creativecommons.org/licenses/by-sa/3.0/deed.en; CC BY 4.0: creativecommons.org/licenses/by/4.0/; CC BY-SA 4.0: creativecommons.org/licenses/by-sa/4.0/

Covers, front matter, backgrounds

front cover Tas van Ommen, *Sleeping Tents for Ice Drilling Field Party at Aurora Basin North Camp*, 2013, © Tas van Ommen; **back cover** Tiarnán Colgan, *View from the Top of Stalker Hill, Looking Northeast Towards the Plateau Near Davis Research Station*, 2021, courtesy Australian Antarctic Division; **endpapers and backgrounds** NASA Goddard Space Flight Center, *Operation IceBridge Turns Five*, 2009, commons.wikimedia.org/wiki/File:Operation_IceBridge_Turns_Five_(15363523929).jpg; Denis Luyten, *Antarctica Sea Ice, Snow Hill*, 2007, commons.wikimedia.org/wiki/File:2007_Snow-Hill-Island_Luyten-De-Hauwere-Sea-Ice-15.jpg; NASA ICE, *Broken Floes*, 2016, commons.wikimedia.org/wiki/File:Broken_floes_(30205786983).jpg; NASA, *Getz Ice Shelf*, 2009, commons.wikimedia.org/wiki/File:Getz_Ice_Shelf.JPG; **ii** Pete Harmsen, *RSV Nuyina Voyage 2*, 2022, courtesy Australian Antarctic Division; **v** John Stoukalo, *Tim Jarvis on the Mawson Expedition*, 2007, courtesy Tim Jarvis; **vi (top to bottom)** Frank Hurley, *Return of the Discovery*, 1931, commons.wikimedia.org/wiki/File:Antarctic_-_Return_of_the_Discovery_in_1931.jpg; Robert Massom, *Drilling for Ice Core Samples*, 2015, courtesy Australian Antarctic Division; Christopher Burns, *Tractor with Ken Done Artwork at the Antarctic Circle Sign*, 2019, courtesy Australian Antarctic Division; Pete Harmsen, *RSV Nuyina Arrives at Casey*, 2022, courtesy Australian Antarctic Division; Frank Hurley, *Frank Hurley, Left, and Sir Ernest Shackleton, Right, in Front of Their Small Tent at Patience or Ocean Camp, Weddell Sea*, 1916, nla.cat-vn2011895; Mark Curran, *Ice Core Drilling Team at Law Dome*, 2008, courtesy Australian Antarctic Division.

Chapter 1: Continent

vii–1 Tas van Ommen, *Snowscape after Recent Snowfall at Law Dome*, 2008, © Tas van Ommen; **2–3** LouieLea, *Navigating among Enormous Icebergs, Including the Largest Ever B-15, Calved from the Ross Ice Shelf of Antarctica*, Shutterstock image 1633144586; **5** Sarah DeWitt, NASA GFSC, *View of a Transantarctic Mountain Glacier from the DC-8*, 2010, https://www.flickr.com/photos/

gsfc/5175667755/; **6** *Australian Antarctic Territory*, courtesy Australian Antarctic Division; **7 (top)** Damien Everett, *Fossilised Southern Elephant Seal Flipper*, 2016, courtesy Australian Antarctic Division; **7 (bottom)** Daniel Eskridge, *Cryolophosaurus Was a Carnivorous Theropod Dinosaur, Known for Its Distinctive Crest, It Lived during the Jurassic in Antarctica*, Shutterstock image 1637654545; **8** Augustus Earle, *Scudding before a Heavy Westerly Gale off the Cape*, 1824, nla.cat-vn2332920; **9** Haughton Forrest, *Tug for Clipper, Tasmania*, c.1860, nla.cat-vn347921; **10–11** Will Standring, *Aurora Australis, Tasmania*, 2015, www.flickr.com/photos/129388046@N04/21926799732, reproduced under CC BY 2.0; **13** Tas van Ommen, *Sleeping Tents for Ice Drilling Field Party at Aurora Basin North Camp*, 2013, © Tas van Ommen.

Chapter 2: Ice

14–15 Joy McCann, *Iceberg*, courtesy Joy McCann; **17** William Hodges (artist) and Benjamin Thomas Pouncy (engraver), *The Ice Islands Seen on the 9th of Jany., 1773*, 1776, nla.cat-vn550672; **18–19** James Wyld, *A New General Chart of the World Exhibiting the Whole of the Discoveries Made by the Late Captain James Cook, F.R.S., with the Tracks of the Ships under His Command*, 1838, nla.cat-vn3792186; **20** Louis Le Breton (artist) and Léon Jean Baptiste (lithographer), *Chasse aux Phoques, le 6 Fevrier 1838*, 1846, nla.cat-vn1469257; **21** John Heaviside Clark (artist) and M. Dubourg (engraver), *Shooting the Harpoon at a Whale*, 1813, nla.cat-vn241439; **22–23** William Duke, *The Flurry*, c.1848, W.L. Crowther Library, Tasmanian Archive and Heritage Office, SD_ILS:130418; **25** Peter Dombrovskis, *Nothofagus cunninghamii, Tasmania*, 1990, nla.cat-vn4980978, courtesy Liz Dombrovskis; **27** Edwin Augustus Porcher, *Van Dieman's Land, Magnetic Observatory at Hobarton*, 1843, nla.cat-vn303365; **29** *Cape Adare, First Landing on the Mainland*, 1895, State Library Victoria, IAN01/04/95/21; **30–31** Clements R. Markham, *Antarctic Exploration. A Plea for a National Expedition*, 1898, nla.cat-vn1640542; **31** Bjørn Christian Törrissen, *A Tribute to the Tasmanian Antarctic Explorer Louis Bernacchi, Erected in the Port of Hobart, Tasmania*, 2008, commons.wikimedia.org/wiki/File:Bernacchi-Statue-In-Hobart-2008.jpg, reproduced under CC BY-SA 3.0; **32** John Watt Beattie, *Expeditionary Member and Sled Dogs on Board the Southern Cross Prior to Leaving for Antarctica, Hobart*, 1898, nla.cat-vn600716; **34 (left)** Louis Bernacchi, *Magnetical Observer, Southern Cross Antarctic Expedition*, 1899, Allport Library and Museum of Fine Arts, Tasmanian Archive and Heritage Office, SD_ILS:616189; **34 (right)** *Bernacchi at Entrance*

Macquarie Island, between 1911 and 1914, nla.cat-vn3158410; **93 (bottom)** Charles A. Sandell, *Wireless Operating Hut, Macquarie Island*, between 1911 and 1914, nla.cat-vn3158132; **94–95** Leslie Russel Blake, *Wireless Hill, the Northern End of Macquarie Island*, between 1911 and 1914, nla.cat-vn3256940; **96** Peter Dombrovskis, *King Penguins, Lusitania Bay, Macquarie Island*, 1984, nla.cat-vn5018699, courtesy Liz Dombrovskis; **99** Frank Hurley, *Penguin and Sea Elephant Life at Macquarie Island*, between 1911 and 1914, nla.cat-vn3244350; **101** Peter Dombrovskis, *Elephant Seals and King Penguins, Macquarie Island*, 1984, nla.cat-vn4738183, courtesy Liz Dombrovskis; **103** Doug Thost, *Light Amphibious Resupply Cargo Vessel Transporting Gear with Stephenson Glacier in Background, Heard Island*, 2004, nla.cat-vn4981335; **104** David Eastman, *Coast of Heard Island, Antarctica*, 1948, nla.cat-vn3765839; **106** *Australians Set up World's Loneliest Weather Post, Antarctica*, 1948, nla.cat-vn7762136; **107** George Brown Goode, *Sketch of Herd's Island: Antarctic Ocean*, 1887, nla.cat-vn6452323; **108–109** David Eastman, *First Party Ashore at Atlas Cove, Heard Island, Antarctica*, 1948, nla.cat-vn3765860; **110–111** David Eastman, *Members of the Heard Island Party Take a Break between Unloading Shifts, Antarctica*, 1948, nla.cat-vn3765917; **112** David Eastman, *British Government Established This Emergency Hut on Heard Island in 1927, Antarctica*, 1948, nla.cat-vn3765780; **113** *Lava Flow on Mawson Peak, Heard Island*, 2012, NASA image GSFC_20171208_Archive_e001509.

Chapter 5: Territory

114–115 Jasmine Nears, *Glacier and Zodiac*, 2017, www.flickr.com/photos/133530809@N08/39129456122, reproduced under CC BY-SA 2.0; **116–117** George Lowe, Australian News and Information Bureau, *Kista Dan Ship Surrounded by Ice Heading towards Sandefjord Bay, Antarctica*, 1955, nla.cat-vn6001492; **118** E.P. Bayliss, *Antarctica*, 1939, nla.cat-vn331525; **121 (top, left)** George Lowe, Australian News and Information Bureau, *Expedition Members Stack Ice outside the Kitchen Door for Drinking Water, Antarctica*, 1955, nla.cat-vn6001490; **121 (top, right)** George Lowe, Australian News and Information Bureau, *Eric Macklin Radio Operator and Bill Storer Postmaster outside the Post Office at Mawson, Antarctica*, 1955, nla.cat-vn6001663; **121 (bottom)** George Lowe, Australian News and Information Bureau, *One of the Huts Demounted and Being Packed for Transport to Mawson Station, Heard Island, Antarctica*, 1955, nla.cat-vn3765870; **123** Pete Harmsen, *Nuyina in Fast Ice*, courtesy Australian Antarctic Division; **125** George Lowe, Australian News and Information Bureau, *A Small Freshwater Lake in the Heart of the Vestfold Hills, Antarctica*, 1955, nla.cat-vn6001662; **126–127** George Lowe, Australian News and Information Bureau, *An Exploration Party with Their Two DUKW Vehicles on a Beach at the Foot*

of a Steep Escarpment of the Vestfold Hills, Antarctica, 1955, nla.cat-vn6001661; **128–129** Felicity Jenkins, *Meteorology Weather Forecaster Neville Martin Releases the Evening Weather Balloon Which Collects Data of Local Atmospheric and Weather Conditions around Davis Station, Antarctica*, 1997, nla.cat-vn796188; **130** Reto Stöckli, *Blue Marble, Eastern Hemisphere*, 2007, NASA image; **132** Alan Campbell-Drury, *Wilkes Handover Ceremony*, 1959, courtesy Australian Antarctic Division; **132–133** Christopher Michael Gregory, *Aerial View of Wilkes Station, Wilkes Land, Antarctica*, 1962, nla.cat-vn5012995; **134** Keith Gooley, *Mike Stracey inside One of the Buildings at Wilkes Station*, 1974, courtesy Keith Gooley; **135** Felicity Jenkins, *Tins of Jam Still in Their Original Boxes Lie Unopened Some Twenty Years after Wilkes Station Had Been Abandoned, near Casey Station, Antarctica*, 1997, nla.cat-vn2317576; **137** Robert Allen, *Casey Tunnel Building*, 1968, courtesy Australian Antarctic Division; **138–139** Graham Denyer, *Casey Station, Australian Antarctic Programme*, 2005, commons.wikimedia.org/wiki/File:Casey_Station_November_2005.jpg, reproduced under CC BY-SA 4.0.

Chapter 6: Station

140–141 Justin Chambers, *Emperor Penguin Huddle*, 2013, courtesy Australian Antarctic Division; **142–143** *John Mayston Béchervaise*, Fryer Library Pictorial Collection, UQFL477,PIC699; **144–145** Phillip Law, *Mawson Station in Summer*, between 1910 and 1962, nla.cat-vn6810869; **147** Frank Hurley, *A Blizzard at Winter Quarters, Cape Denison, Antarctica*, c.1913, nla.cat-vn2386036; **149** Christopher Michael Gregory, *De Havilland DHC-2 Beaver Aircraft and DUKW Military Trucks on Board the Thala Dan, near Wilkes Land, Antarctica*, 1962, nla.cat-vn5013037; **150–151** Jeff Schmaltz, NASA/GSFC, *Mosaic of Antarctica*, 2014, www.flickr.com/photos/gsfc/11822809516; **153** *Norma Ferris, the ABC Radio Presenter of Calling Antarctica, at the ABC studio*, 1969, reproduced by permission of the Australian Broadcasting Corporation–Library Sales; **154** 'Menu du Diner', in *The Home of the Blizzard, Being the Story of the Australasian Expedition 1911–1914* by Douglas Mawson, 1915, nla.cat-vn52882; **155 (top)** Andrew Watson, *West Base Party at Midwinter's Day Dinner*, 1912, Mitchell Library, State Library of New South Wales, ON 144/P21; **155 (bottom)** Frank Hurley, *Mawson and Men at Christmas Dinner with Menu and Balloons, on Board the S.Y. Discovery*, 1930, nla.cat-vn50151; **156–157** Christopher Michel, *Penguin in Antarctica Jumping out of the Water*, 2013, commons.wikimedia.org/wiki/File:Penguin_in_Antarctica_jumping_out_of_the_water.jpg, reproduced under CC BY 2.0; **158** David Eastman, *A Geologist, J.F. Ivanac Makes Friends with a Rock-hopper Penguin, Heard Island, Antarctica*, 1948, nla.cat-vn3765713; **161** *Moyes after His Spell Alone in the Hut*, between 1911 and 1914,

File:Rikr0104 - Flickr_-_NOAA_Photo_Library.jpg;
221 Frank Hurley, *Cecil Thomas Madigan, the Meteorologist, Wearing an Ice-mask, Returning to the Hut during a Blizzard, Australasian Antarctic Expedition*, between 1911 and 1914, nla.cat-vn1614884.

Chapter 9: Wilderness

222–223 U.S. Embassy, *McMurdo Sound*, 2010, www.flickr.com/photos/us_embassy_newzealand/5244733609; **224–225** Grath Wielin, *Ship in Ice, Returning Sun*, 1921, nla.cat-vn1936397; **227** Jan Jansson, *Polus Antarcticus*, 1650, nla.cat-vn2573218; **229** Frank Hurley, *Frank Hurley Photographing from the Tip of the Jib-boom of the Discovery, BANZARE*, c.1930, nla.cat-vn92416; **230** Frank Hurley, *Frank Hurley with Camera on Ice in Front of the Bow of the Trapped Endurance in the Weddell Sea, Shackleton Expedition*, 1915, nla.cat-vn91510; **231** Frank Hurley, *The Endurance Crushed to Death by the Icepacks of the Weddell Sea*, 1915, nla.cat-vn92313; **232–233** Frank Hurley, *The Landing on Elephant Island, Solid Rock Lies beneath Our Feet, This Was Paradise Regained*, 1916, nla.cat-vn489434; **234–235** Frank Hurley, *The Members Who Remained behind at Elephant Island, Shackleton Expedition*, 1916, nla.cat-vn3082159; **236–237** Liam Quinn, *Elephant Island, Antarctica*, 2011, commons.wikimedia.org/wiki/File:Elephant_Island,_Antarctica.jpg, reproduced under CC BY-SA 2.0; **237 (top)** Frank Hurley, *Our Home on Elephant Island Was Built of Two Upturned Boats Laid Side by Side, Twenty Two of Us Lived Like Semi-frozen Sardines within Its Cramped, Dark Interior*, 1916, nla.cat-vn1891176; **237 (bottom)** Frank Hurley, *View of Interior of Hut on Elephant Island*, 1916, nla.cat-vn2011980; **238** Doug Thost, *Mawson's Hut, Commonwealth Bay, Antarctica*, 2001, nla.cat-vn4981324; **239** Frank Hurley, *Interior of Commonwealth Bay Living Hut, Left to Right Mertz, McLean, Madigan, Hunter and Hodgeman*, 1912, nla.cat-vn3122308; **240** John Kissick, *With the 'Aurora' in the Antarctic 1911–1914 By the Captain of the Aurora, John K Davis*, c.1919, nla.cat-vn377359; **241** Frank Hurley, *A Policeman and Two Sailors Inspecting a Dummy Clothed in Antarctic Gear*, 1929, nla.cat-vn783317; **242–243** Christian Stocker, *Antarctica*, 2007, www.flickr.com/photos/chregu/2270556963, reproduced under CC BY-SA 2.0; **245** Sidney Nolan, *Mt Erebus*, 1964, Art Gallery of New South Wales, 1.2011, © The Sidney Nolan Trust. All rights reserved. DACS/Copyright Agency, 2022; **248** *James Castrission and Justin Jones Have Become the First Team to Ski More than 1800km Unsupported in the Antarctic*, 2011, AAP/Hausmann Communications; **248–249** Murray Foubister, *After Dinner, the Boats Headed to the nearby Shore with the Hardy Group that Planned (and Paid Extra for) to Spend the Night in Tents on Shore not Far from Brown Station*, 2015, commons.wikimedia.org/wiki/File:After_dinner,_the_boats_headed_to_the_nearbye_shore_with_the_hardy_group_that_planned_

(and_paid_extra_for)_to_spend_the_night_in_tents_on_shore_not_far_from_Brown_station._(25882566932).jpg, reproduced under CC BY-SA 2.0.

Chapter 10: Environment

250–251 Callan Carpenter, *Killer Whales Hunting a Seal*, 2018, commons.wikimedia.org/wiki/File:Killer_Whales_Hunting_a_Seal.jpg, reproduced under CC BY-SA 4.0; **252** Frank Hurley, *Sedimentary Rock Formation, Australasian Antarctic Expedition*, between 1911 and 1914, nla.cat-vn3208541; **253** Robin Smith, *Geologist at the U.S. Research Station Little America Examining Core Samples from a Drilling Rig Used to Drill the Ice Cap, Bay of Whales, Antarctica*, 1958, nla.cat-vn8067196; **255** Geoff Pryor, *'What Took Them so Long' – Penguins after an Antarctic Oil Spill*, 1989, nla.cat-vn3889893; **257 (top)** Monster4711, *Gondwana Greenpeace Ship*, 1990, commons.wikimedia.org/wiki/File:Gondwana_greenpeace_ship.jpg, reproduced under CC BY-SA 3.0; **257 (bottom)** Andy Loor, *Waste Disposal at US McMurdo Base*, 1987, © Greenpeace/Andy Loor; **258–259** Jason Mundy, *Moss Bed near Old Casey Station*, 2013, courtesy Australian Antarctic Division; **260** Tas van Ommen, *Sleeping Tents at Law Dome with Approaching Storm*, 2004, © Tas van Ommen; **262** Tas van Ommen, *Ice Core Drilling at Law Dome*, 2004, © Tas van Ommen; **263** NASA Goddard MODIS Rapid Response Team, *Satellite View of the Knox, Budd Law Dome, and Sabrina Coasts, Antarctica*, 2011, www.flickr.com/photos/gsfc/6309214790; **264** Tas van Ommen, *Cross-section Slice of an Ice Core Showing Trapped Bubbles of Air from the Atmosphere in the Past*, © Tas van Ommen; **265 (top)** Doug Thost, *Glaciologist Student Shavawn Donoghue Sampling an Ice Core Taken from Brown Glacier, Heard Island, Antarctica*, 2004, nla.cat-vn4981344; **265 (bottom)** Mark Stone, University of Washington, *Close-up of an Ice Core from Antarctica with a Darker Layer of Volcanic Ash*, 2017, www.flickr.com/photos/uwnews/34486867111, reproduced under CC BY 2.0; **267** Stephen Hudson, *Snow Surface at Dome C Station, Antarctica*, 2004, commons.wikimedia.org/wiki/File:AntarcticaDomeCSnow.jpg, reproduced under CC BY-SA 3.0; **268–269** Kathleen, *Iceberg Alley, Antarctica*, 2018, www.flickr.com/photos/72821066@N04/25860128037/, reproduced under CC BY 2.0; **271** Jesse Allen and Robert Simmon, NASA Earth Observatory, *Amery Ice Shelf*, 2012, earthobservatory.nasa.gov/images/77132/amery-ice-shelf.

Acknowledgements

272–273 Christopher Michel, *Emperor Penguins, Gould Bay, Antarctica*, 2013, commons.wikimedia.org/wiki/File:Emperor_Penguins,_Gould_Bay_Antarctica_(16437100992).jpg, reproduced under CC BY 2.0.

Index

The publishing imprint of the
National Library of Australia

Published by NLA Publishing
Canberra   ACT   2600

ISBN: 9781922507334

© National Library of Australia 2022
Text © Joy McCann

Publisher: Lauren Smith
Editors: Amelia Hartney and Melody Lord
Designer: Stan Lamond
Printed in China by Asia Pacific Offset on FSC®-certified paper

Find out more about NLA Publishing at nla.gov.au/national-library-publishing.

Publisher's Note
Many of the people mentioned in this book have made distinguished or notable
contributions in their fields of expertise to knowledge, practice or society,
achievements that have been recognised in later life with formal honours and
academic qualifications. For ease of reading, we have not included honorifics
and include academic titles only at first mention.

The National Library of Australia would like to thank the Australian Antarctic
Division (AAD) for its significant contribution to this book.

Excerpts from Douglas Mawson's *The Home of the Blizzard* courtesy the Trustees
of the Estate of the Late Sir Douglas Mawson, Alun Thomas and Andrew McEwin;
other Douglas Mawson excerpts (p.241, p.100, p.54) courtesy the South Australian
Museum; quote about John Béchervaise on p.142 © Peter Keague/*The Independent*;
quote about Dick Smith's Antarctica fly-over on p.247 courtesy *The Sydney Morning
Herald*; quote from Dr Geoff Moseley on p.254 courtesy *The Canberra Times*.

A catalogue record for this
book is available from the
National Library of Australia